SUSTAINABILITY FOR THE NATION

Resource Connections and Governance Linkages

Committee on Sustainability Linkages in the Federal Government

Science and Technology for Sustainability Program

Policy and Global Affairs Division

NATIONAL RESEARCH COUNCIL
OF THE NATIONAL ACADEMIES

THE NATIONAL ACADEMIES PRESS
Washington, D.C.
www.nap.edu

THE NATIONAL ACADEMIES PRESS 500 Fifth Street, NW Washington, DC 20001

NOTICE: The project that is the subject of this report was approved by the Governing Board of the National Research Council, whose members are drawn from the councils of the National Academy of Sciences, the National Academy of Engineering, and the Institute of Medicine. The members of the committee responsible for the report were chosen for their special competences and with regard for appropriate balance.

This study was supported by a grant from the U.S. Environmental Protection Agency under award number EP10H002179, U.S. Geological Survey under award number G10AP00149, U.S. Department of Energy under award number DE-PI0000010, TO #8, National Aeronautics and Space Administration under award number NNX11AB43G, U.S. Department of Agriculture under award number 59-0790-1-124, National Science Foundation under award number CBET-1135117, National Oceanic and Atmospheric Administration under award number WC133R-11-CQ-0048, TO #1, BP, Lockheed Martin, the David and Lucile Packard Foundation under award number 2011-36690, and the Cynthia and George Mitchell Foundation. Any opinions, findings, conclusions, or recommendations expressed in this publication are those of the author(s) and do not necessarily reflect the views of the organizations or agencies that provided support for the project.

International Standard Book Number-13: 978-0-309-26230-9
International Standard Book Number-10: 0-309-26230-5

Additional copies of this report are available for sale from the National Academies Press, 500 Fifth Street, NW, Keck 360, Washington, DC 20001; (800) 624-6242 or (202) 334-3313; http://www.nap.edu.

Printed in the United States of America.

THE NATIONAL ACADEMIES
Advisers to the Nation on Science, Engineering, and Medicine

The **National Academy of Sciences** is a private, nonprofit, self-perpetuating society of distinguished scholars engaged in scientific and engineering research, dedicated to the furtherance of science and technology and to their use for the general welfare. Upon the authority of the charter granted to it by the Congress in 1863, the Academy has a mandate that requires it to advise the federal government on scientific and technical matters. Dr. Ralph J. Cicerone is president of the National Academy of Sciences.

The **National Academy of Engineering** was established in 1964, under the charter of the National Academy of Sciences, as a parallel organization of outstanding engineers. It is autonomous in its administration and in the selection of its members, sharing with the National Academy of Sciences the responsibility for advising the federal government. The National Academy of Engineering also sponsors engineering programs aimed at meeting national needs, encourages education and research, and recognizes the superior achievements of engineers. Dr. Charles M. Vest is president of the National Academy of Engineering.

The **Institute of Medicine** was established in 1970 by the National Academy of Sciences to secure the services of eminent members of appropriate professions in the examination of policy matters pertaining to the health of the public. The Institute acts under the responsibility given to the National Academy of Sciences by its congressional charter to be an adviser to the federal government and, upon its own initiative, to identify issues of medical care, research, and education. Dr. Harvey V. Fineberg is president of the Institute of Medicine.

The **National Research Council** was organized by the National Academy of Sciences in 1916 to associate the broad community of science and technology with the Academy's purposes of furthering knowledge and advising the federal government. Functioning in accordance with general policies determined by the Academy, the Council has become the principal operating agency of both the National Academy of Sciences and the National Academy of Engineering in providing services to the government, the public, and the scientific and engineering communities. The Council is administered jointly by both Academies and the Institute of Medicine. Dr. Ralph J. Cicerone and Dr. Charles M. Vest are chair and vice chair, respectively, of the National Research Council.

www.national-academies.org

Preface

In July 2011, the National Research Council's (NRC's) Science and Technology for Sustainability Program initiated a new study, *Sustainability Linkages in the Federal Government.* This followed a series of discussions held by the NRC's Roundtable on Science and Technology for Sustainability that explored linkages between topics critical to long-term sustainability. The premise is that achieving sustainability is a systems challenge that cannot be addressed by separately optimizing pieces of the system. To address this systems challenge, an ad hoc committee with a wide range of expertise and experience in government, academia, and business was convened. Brief biographies of the individual committee members are provided in Appendix A. The committee was charged to produce a report with consensus findings that provides an analytical framework for decision formulation and decision making related to linkages of sustainability. This framework can be used by U.S. policy makers and regulators to assess the consequences, trade-offs, and synergies of policy issues involving a systems approach to long-term sustainability and decisions on sustainability-oriented programs. The framework was to include social, economic, and environmental domains of sustainability, highlighting certain dimensions that are sometimes left unaccounted for in cross-media analyses.

During the course of the study, the committee conducted several fact-finding meetings and committee meetings. The first committee meeting was held September 20-21, 2011, in Washington, D.C. During this meeting, sponsors discussed areas of interest to their agency or organization, and several panel discussions addressed a variety of perspectives (state and local, industry, non-governmental, and national) on sustainability linkages.

Three subsequent fact-finding meetings explored specific examples that cut across a variety of geographies and scales and featured a range of sustainability challenges. The purpose of these meetings was to examine in detail a number of approaches to challenges involving either connections among sustainability-related resources, or of linkages across agencies addressing such challenges. At these meetings, the committee heard from and questioned those involved about issues of science, monitoring, organization, communication, and governance. The geographies included sustainability management of coastal systems (the

Puget Sound, the Great Lakes; committee meeting in Seattle, WA, February 6-8, 2012), regional nonurban systems (the Mojave Desert, the Platte River; committee meeting in Omaha, NE, April 11-12, 2012), and urban systems (Phoenix, Philadelphia; committee meeting in Tempe, AZ, June 11-12, 2012). The sustainability issues that were the focus of the examples involved the tension of managing multiple users of a given resource, and multiple stressors on elements of these systems. System elements included water quality and availability, ecosystem health, endangered species, energy, transportation, urban infrastructure, public health, commerce, and food production. The linkages involved federal agencies, regional organizations, state agencies, local government, nongovernmental organizations, and citizen stakeholder groups. A fifth committee meeting was held on July 16-19, 2012, in Woods Hole, MA, to begin to draft the consensus report, and a sixth meeting was held on October 11-12, 2012, in Washington, DC. An agenda for each meeting is provided in Appendix C.

The committee gratefully acknowledges all of the speakers for their informative presentations, and Derek Vollmer, National University of Singapore, and Stephanie Ariganello, Michigan Sea Grant, for preparing background papers for the meetings. The information provided at these meetings is used throughout this report and provided important perspectives which were utilized in this report's findings and conclusions.

In this report, Chapter 1 first describes the challenge that the committee addressed. Chapter 2 discusses the current impediments to effective government action, Chapter 3 explores the fact-finding examples and the lessons they provide, Chapter 4 develops the decision framework for linkage challenges, and Chapter 5 provides a vision for improved responses to sustainability linkages. The committee acknowledges Janene Cowan, University of Minnesota, and Kurt Barnes, Barnes Bros., for providing support for visual materials for the decision framework.

The report would not have been possible without the sponsors of this study, including the U.S. Environmental Protection Agency, U.S. Geological Survey, U.S. Department of Energy, National Aeronautics and Space Administration, U.S. Department of Agriculture, National Science Foundation, National Oceanic and Atmospheric Administration, BP, Lockheed Martin, the David and Lucile Packard Foundation, and the Cynthia and George Mitchell Foundation. The April 2010 planning meeting was supported by the Interface Environmental Foundation.

On behalf of the committee, I want to express our thanks and appreciation to Marina Moses, director of the Science and Technology for Sustainability Program, Jennifer Saunders, the program officer responsible for our study, and Emi Kameyama, program associate, for the time and effort they put into assembling the committee, planning the meetings, and organizing the report. I also thank the National Academies staff, Dominic Brose, program officer; Dylan Richmond, research assistant; Patricia Koshel, senior program officer; Sara Frueh, media officer II; Stephen Mautner, executive editor; Adriana Courembis, financial associate; Radiah Rose, editorial projects coordinator; and Kathleen McAllister, associate program officer (through October 2010), for their support and assistance with study activities.

Finally, I thank, especially, the members of the committee for their tireless efforts throughout the development of this report.

Thomas Graedel, *Chair*
Committee on Sustainability Linkages
in the Federal Government

Acknowledgments

This report has been reviewed in draft form by individuals chosen for their diverse perspectives and technical expertise, in accordance with procedures approved by the National Academies' Report Review Committee. The purpose of this independent review is to provide candid and critical comments that will assist the institution in making its published report as sound as possible and to ensure that the report meets institutional standards for objectivity, evidence, and responsiveness to the study charge. The review comments and draft manuscript remain confidential to protect the integrity of the process.

We wish to thank the following individuals for their review of this report: David Allen, University of Texas, Austin; Craig Benson, University of Wisconsin-Madison; Nancy Creamer, North Carolina State University; Kirk Emerson, University of Arizona; Vasilis Fthenakis, Columbia University; Shelley Hearne, Johns Hopkins University; John Onderdonk, California Institute of Technology; and Kenneth Ruffing, Independent Consultant.

Although the reviewers listed above have provided many constructive comments and suggestions, they were not asked to endorse the conclusions or recommendations, nor did they see the final draft of the report before its release. The review of this report was overseen by Robert Frosch, Harvard University and Richard Wright, National Institute of Standards and Technology (Retired). Appointed by the National Academies, they were responsible for making certain that an independent examination of this report was carried out in accordance with institutional procedures and that all review comments were carefully considered. Responsibility for the final content of this report rests entirely with the authoring committee and the institution.

Contents

BOXES AND FIGURES

BOXES

FIGURES

Abbreviations and Acronyms

ASU	Arizona State University
BLM	Bureau of Land Management, U.S. Department of the Interior (DOI)
BOR	Bureau of Reclamation
BPA	Bonneville Power Administration
CAP LTER	Central Arizona-Phoenix Long-Term Ecological Research
CDC	Centers for Disease Control and Prevention
CDFG	California Department of Fish and Game
CEC	California Energy Commission
CEF	Corporate Eco Forum
CEQ	Council on Environmental Quality
CSO	Combined Sewer Overflow
DMG	Desert Managers Group
DOD	U.S. Department of Defense
DOE	U.S. Department of Energy
DOI	U.S. Department of the Interior
DOT	U.S. Department of Transportation
DRECP	Desert Renewable Energy Conservation Plan
EPA	U.S. Environmental Protection Agency
ESA	Federal Endangered Species Act
FEMA	Federal Emergency Management Agency
FWS	U.S. Fish and Wildlife Service
GAO	U.S. Government Accountability Office
GC	Governance Committee (Platte River Program)
GPRA	Government Performance and Results Act
GSI	Green Stormwater Infrastructure
HUD	U.S. Department of Housing and Urban Development
IJC	International Joint Commission
JLARC	State of Washington Joint Legislative Audit and Review Committee
LTER	National Science Foundation's Long-Term Ecological Research
NASA	National Aeronautics and Space Administration
NCCP	Natural Communities Conservation Plan
NEPA	National Environmental Protection Act of 1969

NGO	Nongovernmental Organization
NIH	National Institutes of Health
NOAA	National Oceanic and Atmospheric Administration
NOP	National Ocean Policy
NPS	National Park Service, DOI
NRC	National Research Council
NSF	National Science Foundation
OECD	Organisation for Economic Co-operation and Development
OMB	Office of Management and Budget
PHS	Pennsylvania Horticultural Society
PMF	Presidential Management Fellow
PSP	Puget Sound Partnership
PWD	Philadelphia Water Department
REAT	Renewable Energy Action Team
SEES	Science, Engineering, and Education for Sustainability Program, National Science Foundation
SES	Senior Executive Service
USDA	U.S. Department of Agriculture
USGS	U.S. Geological Survey, DOI
WSAS	Washington State Academy of Sciences

Summary

A "sustainable society," according to one definition, "is one that can persist over generations; one that is far-seeing enough, flexible enough, and wise enough not to undermine either its physical or its social system of support."[1] This definition is consistent with the intent of the statement in the National Environmental Protection Act of 1969 (NEPA): "To create and maintain conditions under which humans and nature can exist in productive harmony and that permit fulfilling social, economic, and other requirements of present and future generations." Sustainability issues occur at all scales from the global, such as the challenge of meeting the needs of a potential global population of 9 billion, to the national scale, to the regional and local scales.

In their efforts to ensure sufficient fresh water, food, energy, housing, health, and education while maintaining ecosystems and biodiversity for future generations, federal agencies discover that, for a variety of reasons, they are not well organized to address the crosscutting nature of sustainability challenges. Moreover, those crosscuts are often the crucial points where progress can be made, and generally involve agencies and organizations at levels other than federal. In some instances, it is difficult to get all stakeholders to the table, and challenges persist. In others, collaborative approaches of various kinds succeed in surmounting the challenges and making good progress toward important achievements. Two examples are useful in demonstrating these alternative outcomes.

In the 1990s many agencies and individuals in the Seattle area recognized that the ecological health of Puget Sound, its prospects for a continuing shellfish industry, and its attractiveness as a recreational resource were under threat. They came together to focus on that challenge and have modestly improved water quality and shellfish survival in Puget Sound. However, agencies making local land-use decisions affecting the Sound have declined to be involved, and progress in addressing Puget Sound's problems continues to be impeded by runoff from poorly located land developments.

---

[1]Meadows, D. H., D. L. Meadows, and J. Randers. 1992. Beyond the Limits. White River Junction, VT: Chelsea Green Publishing.

1

Also in the 1990s, concerns were raised about the maintenance of endangered species, energy generation, agriculture, and recreation in the central Platte River of Nebraska. After considerable discussion, agencies of the federal government, three states, power providers, water managers, and others came together to create a shared vision and to establish responsibility for sustainable management of the central Platte River. This shared vision has led to improved environments for endangered species, better collaborative water management, and more stable hydropower production.

Why did these two situations challenge established governance systems? Why did one approach succeed, but not the other? The answer to these questions relates to systems thinking and to the challenges that arise when traditional approaches to governance meet the need for systems thinking. The legendary ecologist John Muir wrote in 1911 that "when we try to pick out anything by itself, we find it hitched to everything else in the Universe."[2] His perceptive statement applies to water, land, wildlife, and other aspects of the natural world, as well as to the interactions that link humans and nature. Many decades later, it has become increasingly obvious that the statement is also relevant to resource governance.

To explore how such sustainability challenges might be better addressed, a committee with a wide range of expertise and experience in government, academia, and business was convened by the National Research Council (NRC) to provide guidance on issues related to sustainability linkages in the federal government. This report is the result of the committee's investigations and deliberations.

The committee was charged to produce a report with consensus findings that provides an analytical framework for decision making related to linkages of sustainability. This framework can be used by U.S. policy makers and regulators to assess the consequences, tradeoffs, and synergies of policy issues involving a systems approach to long-term sustainability and decisions on sustainability-oriented programs. The framework is to include social, economic and environmental domains of sustainability, highlighting certain dimensions that are sometimes left unaccounted for in cross-media analyses. The committee was also asked to:

- identify impediments to interdisciplinary, cross-media federal programs;
- recommend priority areas for interagency cooperation on specific sustainability challenges; and
- highlight scientific research gaps as they relate to these interdisciplinary, cross-media approaches to sustainability.

To address this statement of task, the committee convened a series of fact-finding meetings, commissioned expert-authored examples, and reviewed the pertinent literature, as discussed below in more detail.

---

[2]Muir, J. 1911. My First Summer in the Sierra. Boston, MA: Houghton Mifflin.

RESOURCE CONNECTIONS AND GOVERNANCE LINKAGES

This report focuses on government efforts but is grounded in systems thinking, incorporating social, economic, and environmental considerations; for clarity's sake, it uses distinct terms for the *connections* across social-ecological systems and governance *linkages* that are needed for successful management of connected systems. For example, managing water resources sustainably means considering not just water quality and quantity, but also its *connection* to air quality, land use that may involve food and energy production or urban development, drinking water security, electricity from hydropower, fisheries, recreation, and impacts on human health.[3] Governing for sustainability requires bringing the right complement of people to the table, acknowledging the *linkages* across societal and governance institutions. Thus, managing water resources would require enabling and facilitating effective linkages among dozens of federal, state, local and sometimes international institutions and organizations. Governance by its traditional nature (and often by statute) is compartmentalized, but maximizing one variable at a time is a path to suboptimal results. Better results are likely to be achieved by managing the connections and optimizing governance linkages.

The systems that must be considered in addressing sustainability challenges are referred to in this report as social-ecological systems. These complex systems include the natural resource domains (air, fresh water, coastal oceans, land, forests, soil, etc.), built environments (urban infrastructure such as drinking water and wastewater systems, transportation systems, energy systems), and the social aspects of complex human systems (public health, economic prosperity, and the like).

Figure S-1 provides a graphical depiction of the challenge, where key resource domains, including water, land, energy, and nonrenewable resources, are shown as squares, and areas that require these resources (industry, agriculture, nature, and domestic) are depicted as ovals. Human health and well-being interacts with all of these. It is common that scientists and decision makers specialize in one of these topics and are relatively unaware of the important constraints that may occur as a result of inherent connections with other topics. As the diagram demonstrates, a near-complete connection exists between all of these domains, even though tradition and specialization encourage a focus on only one part of these highly interconnected systems.[4]

---

[3]The Organisation for Economic Co-operation and Development (OECD) Environmental Strategy refers to this as "integrated water resources management." See OECD. 2012. Review of the Implementation of the OECD Environmental Strategy for the First Decade of the 21st Century: Making Green Growth Deliver. Meeting of the Environment Policy Committee (EPOC) at Ministerial Level, March 29-30, 2012. Online. Available at http://www.oecd.org/env/50032165.pdf. Accessed February 13, 2013.

[4]Graedel, T. E., and E. van der Voet. 2010. Linkages of sustainability: An introduction. Pp. 1-10 in Linkages of Sustainability, T. E. Graedel and E. van der Voet, eds. Cambridge, MA: MIT Press.

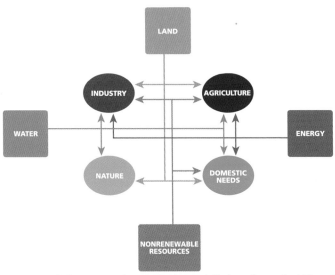

FIGURE S-1 The links among the needs for and limits of sustainability. SOURCE: Graedel, T. E., and E. van der Voet, 2010, adapted from Figure 1.2 The links among the needs for and limits of sustainability. Reprinted with permission from the MIT Press.

While it may be challenging to address connections across natural and human system domains, successful governance requires it. Ignoring connections raises the risks of unintended consequences from policy actions and can result in ineffective and inefficient outcomes. Sustainable management of connected systems calls for governance that effectively links across domains, as well as across geographic and temporal scales.

DECISION FRAMEWORK

Effective governance for sustainability requires strong organizational interaction and collaboration. A number of impediments or barriers frustrate federal government efforts to create linkages to address sustainability issues. These include legal limitations in the form of structural or vertical fragmentation of authority; funding mechanisms that favor short-term, single-agency initiatives rather than longer-term cross-agency projects; a lack of access to or coordination of such foundational elements as research and information/data; and the culture of government. There are very few institutional bridges, practices, or processes that incentivize building and sustaining the necessary linkages. The difficulties of creating or forging such ties were evident in many of the committee's fact-finding examples, as were the ways in which such impediments or barriers could be overcome. Because many sustainability issues cross agency boundaries and require long-term investment, these situations create challenges to effective government response.

However, steps can be taken to reduce these barriers and create structures and incentives for greater collaboration where it is needed or beneficial, as well as to engage relevant decision makers and experts to develop and implement solutions. The optimum approach to doing so is to follow a structured decision framework that reflects relevant connections, identifies those in charge and those affected, and surveys what can be done to integrate the needs and responsibilities of all. Figure S-2 presents a graphic representation of the decision framework recommended by the committee. The purpose of this framework is to lay out a structured but flexible process from problem formulation through achievement of measureable outcomes, a process that engages agencies and stakeholders in goal-setting, planning, knowledge building, implementation, assessment, and decision adjustments. It is designed to be used in addressing place-based sustainability challenges as well as in policy formulation and rule-making. The framework incorporates an iterative (or incremental) process that can yield solutions to a wide range of issues that vary in scope, characteristics, and time.

The framework is depicted in four phases: (1) preparation and planning; (2) design and implementation; (3) evaluation and adaptation; and (4) long-term outcomes. It is meant to apply to the creation of a sustainability program and projects. A brief description of each of the framework phases is given below, and detailed information about each of the phases can be found in Chapter 4.

Phase 1: Preparation and Planning. This phase has three major steps that need to occur prior to the actual program or project design: (1) frame the problem (determine baseline conditions, key drivers, metrics, and goals based on these metrics); (2) identify and enlist partners; and (3) develop a project management plan. This important phase and its associated steps are often overlooked or done in an incomplete or piecemeal fashion.

Phase 2: Design and Implementation. This phase has three main steps, including: (1) define goals; (2) design action plan; and (3) implement plan.

Phase 3: Evaluation and Adaptation. This phase focuses on realizing short-term outcomes, assessing outcomes, and adjusting actions. Outcomes are assessed and evaluated relative to the baseline conditions established in Phase 1.

Phase 4: Long-Term Outcomes. Long-term outcomes are on the scale of several years or more and should closely track the goals identified in the first phase. While performance is assessed and adjustments continue to be made during this phase, as in the previous one, a point is reached where a formal assessment is needed. Using outcome measures developed under Phase 2, at this stage evaluations are conducted to see if short- and long-term outcomes are meeting goals. Ideally, this evaluation should be able to be compared to the baseline evaluation finalized in Phase 2. Based on this evaluation, necessary changes to the team, goals, outcomes and measures, management plans, design, implementation, or maintenance are made.

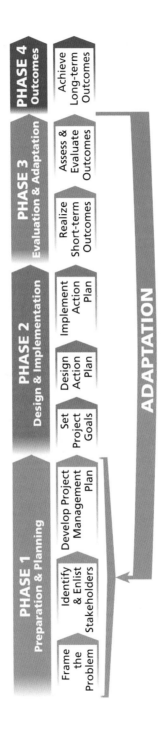

FIGURE S-2 The committee's proposed decision framework.

When well executed, this framework process will enhance legitimacy, encourage systems thinking and the relevance of government actions, and may also result in streamlined and more efficient governance. An additional benefit is that the experiences and lessons learned in applying this process are fed back to the participating organizations and individuals, improving both future efforts and, potentially, government efficiency.

PRIORITY DOMAINS AND ISSUE AREAS

The committee was charged with prioritizing sustainability issues that present significant connections among resource domains and across economic, social, and environmental dimensions. Using several criteria, the committee has identified priority issues below that would benefit from the processes envisioned by the decision framework. The criteria include issues that are nationally important, require interdisciplinary data and analysis, involve multiple interconnected resource domains, would benefit from greater coordination, have the potential to leverage nongovernmental knowledge and resources, and would result in positive returns on investment. Opportunities to better identify and address sustainability linkages are extensive.[5] The committee applied the selection criteria to highlight several significant issue clusters below.

- **Connections among energy, food, and water:** The availability and abundance of affordable supplies of energy, food, and water are vital to sustaining healthy populations and economic prosperity.
- **Diverse and healthy ecosystems:** Ecosystems and their components and functions provide "services" to human communities—for example, in terms of water supplies and quality, coastal storm buffers, productive fisheries, pollination, air pollution absorption, and soil quality along with many extractive and other uses of resources.
- **Enhancing resilience of communities to extreme events:** There is a significant need to assess infrastructure and community vulnerabilities to natural and human-caused disasters and to develop more coordinated strategies for addressing them.
- **Human health and well-being:** Clean air and water, nutritious food, regular physical activity, and protection from toxic exposures and injuries are among the requirements for human health and well-being; each of these is affected by sustainability initiatives.

---

[5]A 1999 NRC report identifies eight priority areas needing greater attention and coordinated efforts to enhance sustainable outcomes that meet economic, social, and environmental goals. NRC. 1999. Our Common Journey. Washington, DC: National Academies Press.

FINDINGS

Through its review of the literature, fact-finding examples (Box S-1), and expert judgment, the committee has arrived at the following findings:

Sustainability issues inherently involve connections among environmental, economic, and social issues, making them extraordinarily difficult to address on their own terms (Chapter 1). Connections among these realms can vary by scope, scale, and time.

The federal government is generally not organized or operated to deal with this complexity (Chapter 2). Agencies have distinct missions, for the most part focusing on one arena (e.g., health, energy, environment), with programs addressing one exigency (e.g., natural disaster, statute), in one domain (e.g., air, water, land use), and one time frame (e.g., dictated in statute, term of office). Although the complexity of sustainability issues typically exceeds the scope of any single agency, there are structural and cultural impediments to agencies' working together. The paucity of cross-government mechanisms to facilitate sustainability constitutes a barrier. The pejorative, but nevertheless accurate, description for this fragmentation of authority is the stovepipe or silo effect: Each agency focuses on implementing its own statutory mandate.[6]

Collaboration, network governance, and other forms of multiauthority initiatives are more effective and have greater durability when supported by some form of legal status that comes from legislation, executive orders, etc. With or without such legal status, however, **there are models of collaborative networks and shared governance that transcend organizational and resource boundaries (Chapter 3).** Such efforts are often the product of individuals who are not given incentives to do so, but who believe in an issue and have found ways to work that are not disallowed. In other cases, success results from action taken by leadership in the absence of definitive responsibility. Some successes result from modest efforts, while others are a product of comprehensive changes. Either staff or leadership can initiate a process leading to success; however, both ultimately need to be involved.

Success on sustainability issues in the federal government depends upon several key factors: engaging stakeholders throughout the process; including and integrating environmental, economic, and social dimensions; using a strong science base and processes that link science and decision making; and reaching stakeholder agreement on the nature of important connections (Chapter 4).

---

[6]Kettl, D. F. 2002. The Transformation of Governance: Public Administration for Twenty-first Century America. Baltimore, MD: Johns Hopkins University Press.

BOX S-1
Summary of Fact-Finding Examples

Philadelphia, an older city of 1.5 million people, is one of the poorest large cities in the nation, with deteriorating infrastructure that needs to be replaced to meet federal standards. Sustainability was a major issue in a recent mayoral election and the city has since adopted Greenworks Philadelphia, a plan to make the city more sustainable by 2015. Innovative initiatives such as converting vacant city lots into parks and the Green Stormwater Initiative have expanded the physical green footprint of the city while reducing crime and stress among residents.

Phoenix, a rapidly growing and ethnically diverse desert city, faces a unique combination of sustainability challenges, including water scarcity, poor urban air quality, significant loss of biodiversity, increasing demands on energy resources, and urban heat island effects on public health. A plan released by then-Mayor Phil Gordon in 2009 calls for Phoenix to become the country's first carbon-neutral city, and it has strong linkages to state and national groups, local communities, and corporations to help achieve this goal; however, the mayor's office has only one person focused on sustainability, with no designated budget authority, and its desert locale means that the region lacks key natural resources such as water.

The Platte River plays a key role in providing electricity and irrigation to residents of Nebraska, Wyoming, and Colorado; however, these uses have diminished the water supply in the central Platte River, a critical habitat for four endangered species. A cooperative agreement between the states and the federal government has established a shared vision and responsibility for managing the central portion of the river to protect these endangered species, while allowing for power generation and irrigation on the upper portions.

The Mojave Desert is a fragile ecosystem that has experienced much development over the past few years, from solar- and wind-energy development to tracts of land used for mining, agriculture, housing, and military training, but tension exists between active uses of the land and a desire to preserve species habitat. Numerous agencies and organizations oversee land in the region and have various uses for it, making it imperative that government agencies cooperate to protect the desert's resources while managing public use and supporting agency missions.

The Great Lakes of North America, the largest body of fresh water on the planet, have played a critical role historically, environmentally, economically, and culturally. Yet the Great Lakes Basin is administratively challenging, spanning the U.S. and Canada and containing numerous states, provinces, native peoples, and local governments. Severe water quality problems, past and present, have led to many agreements at multiple scales, and regional governance institutions, such as the International Joint Commission (IJC), have arisen to combat challenges faced by the Great Lakes Basin.

The Pacific Northwest contains multiple environmentally sensitive regions that have required responses across several governance levels. The Columbia River has numerous federal dams providing hydroelectricity to the region, but this has severely impacted the ability of salmon to reach their traditional spawning grounds, requiring local, state, and federal governments to work together to optimize the river's capacity to meet varying demands. Because urbanization in the Seattle-Tacoma metropolitan region has taken a toll on the Puget Sound, the State of Washington created a new agency, the Puget Sound Partnership, to environmentally protect and restore the Sound and partner with other agencies such as the U.S. EPA to help achieve this goal.

Sustainability management, when effectively implemented, creates greater value, minimizes unintended consequences, and ultimately improves the efficiency of government activities (Chapters 1 and 4). Indeed, it is likely that applying a sustainability decision framework and coordinating the work of multiple agencies and the public and private sectors will result in better outcomes while conserving resources and effort.

RECOMMENDATIONS

To address these findings, the committee makes several recommendations to guide federal agencies in effectively addressing sustainability issues.

Federal agencies should adopt or adapt the decision framework described above (and developed in greater detail in Chapter 4). Special attention should be paid to 1) incorporating adaptive management[7] approaches, 2) engaging all stakeholders, including state and local governments and nongovernmental organizations (NGOs), through iterative processes to the extent possible, and 3) communicating both objectives and progress toward those objectives throughout the process to all concerned. To maximize the potential for success, several additional elements must also be in place. Perhaps most important is to build sustainability into the fabric of an agency or organization: in its mission statement, its goals and objectives, and its organizational and management structure. Also very important are structuring sustainability decision making on long time frames and assessing ways to maximize benefits in all sustainability solutions and approaches.

A National Sustainability Policy should be developed that will provide clear guidance to the executive agencies on addressing governance linkages on complex sustainability problems and inform national policy on sustainability (Chapter 5). A process should be established for developing this policy, as well as a strategy for implementing it. All stakeholders, including the private sector and NGOs, should be provided an opportunity for contributing to this process. Once the policy is in place, agencies should develop specific plans to define how they expect to implement the policy. In implementing the National Sustainability Policy, consideration should be given to the creation of open and transparent oversight involving the public, state legislatures, Congress, and the President.

---

[7]The U.S. Geological Survey (USGS) describes adaptive management "as a systematic approach for improving natural resource management, with an emphasis on learning about management outcomes and incorporating what is learned into ongoing management. Adaptive management can be viewed as a special case of structured decision making, which deals with an important subset of decision problems for which recurrent decisions are needed and uncertainty about management impacts is high." USGS. 2012. Adaptive Management. Online. Available at http://www.usgs.gov/sdc/adaptive_mgmt. html. Accessed January 26, 2012.

Agencies should support innovations in efforts to address sustainability issues by identifying key administrative, programmatic, funding, and other barriers and by developing ways to reduce these barriers (see Chapter 2). Agencies need not await structural overhauls to strengthen their capacity to address sustainability issues. Agencies can begin by preparing a high-level systems map illustrating key connections and linkages, which can then be deployed widely across federal agencies to encourage policy coordination for any sustainability-related program or project.

Agencies should legitimize and reward the activities of individuals who engage in initiatives that "cross silos" in the interest of sustainability, both at the staff and leadership level (Chapter 5). Among other things, agencies should develop personnel performance measures that emphasize collaboration and the design and implementation of interagency, integrated approaches to addressing sustainability issues. Agencies should nurture "change agents" both in the field and at regional and national offices, an effort that may include revisions to managers' performance plans, rewards, and training as well as better alignment of policy tools to support collaboration. Similarly, agencies should encourage and enable cross-agency management and funding of linked sustainability activities. In some cases, statutory authority to cross silos as well as to develop cross-agency funding on integrated cross-domain issues may be required.

Agencies should support long-term, interdisciplinary research underpinning sustainability (Chapter 5). Among other things, the committee recommends funding robust research to provide the scientific basis for sustainability decision making. Sustainability challenges play out over long time scales; therefore, agencies should invest in long-term research projects on time scales of decades to provide the necessary fundamental scientific understanding of sustainability. An example of such a long-term research program is the National Science Foundation's (NSF's) Long-Term Ecological Research Program (LTER). Moreover, successfully meeting sustainability challenges requires that agencies support additional interdisciplinary, cross-program research, such as NSF's Science, Engineering, and Education for Sustainability Program (SEES). Although the impact of sustainability on human well-being is critically important, scientific information on this relationship is woefully inadequate and incomplete and needs to be strengthened at major health funding agencies, such as the National Institutes of Health (NIH). The committee also recommends a systematic analysis of network and governance models and adaptive decision making efforts to identify common issues and challenges.

Federal agencies that support scientific research should be incentivized to collaborate on sustained, cross-agency research (Chapter 5). Sustainability should be supported by a broader spectrum of federal agencies, and additional federal partners should become engaged in science for sustainability. Federal agencies should collaborate in designing and implementing cross-agency research portfolios to better leverage funding.

It will also be critical to develop training for leadership and staff that includes both scientific and management aspects of sustainability issues and that addresses the system and agency linkages needed to achieve sustainability outcomes. Similar training should be incorporated into entry-level programs such as the Presidential Management Fellows (PMF) program and into senior-level training such as the Senior Executive Service (SES) program.

Chapter 1

The Challenge of Managing Connected Systems

C. S. Lewis wrote, "Everything connects with everything else, but not all things are connected by the short and straight roads we expected" (Lewis, 1947). Those who hope to meet the challenges of providing sufficient fresh water, food, energy, housing, health, and education to the world's 9 billion people while maintaining ecosystems and biodiversity for future generations know Lewis was correct on both counts.

WHAT IS SUSTAINABILITY?

A "sustainable society," according to one definition, "is one that can persist over generations; one that is far-seeing enough, flexible enough, and wise enough not to undermine either its physical or its social system of support" (Meadows et al., 1992). This definition is consistent with the intent of the statement in the National Environmental Protection Act of 1969 (NEPA): "To create and maintain conditions under which humans and nature can exist in productive harmony and that permit fulfilling social, economic, and other requirements of present and future generations." Sustainability issues occur at all scales from the global, such as the challenge of meeting the needs of a potential global population of 9 billion, to the national scale, to the regional and local scales.

Among many other disciplines, science plays a key role in advancing sustainability. Key features of the emerging field of sustainability science, launched just after the turn of the current century (Kates et al., 2001), include that it is problem driven; focuses on dynamic interactions between nature and society; and requires an integrated understanding of complex problems, necessitating a transdisciplinary, systems-based approach (see Box 1-1 for more information about important elements of the approach to sustainability).

A central goal of sustainability, although one often overlooked in this context, is to maintain and enhance human well-being. Human well-being is a multidimensional concept that includes physical and mental health across the lifespan, from prenatal development to old age. It also includes happiness, a more elusive state of being that has been increasingly studied and quantified in recent years. Issues of equity and security are other important dimensions of well-being, and range from safe neighborhoods to secure employment to the ability to pay for food and utilities to peace and security at the national level. Finally, well-being extends across generations; people who know that their children and grandchildren will have the opportunity for good lives enjoy an added measure of well-being. Government plays an important role in creating a sense of well-being; well-being is enhanced when society believes government is functioning in an efficient and effective manner.

A common and useful way of thinking about sustainability is to refer to the three overlapping domains of sustainability. Each domain—environment, social, economic—contributes essential components to sustain human well-being (Figure 1-1).

BOX 1-1
The Sustainability Approach

Key features of the sustainability approach include: its "problem-driven" quality, an orientation toward generating and applying knowledge that supports decision making for sustainability; its focus on dynamic interactions between nature and society, using the framework of complex socioeconomic-ecological (also called human-environment) systems (Gunderson and Holling, 2002); its goal of an integrated understanding of complex problems, requiring trans-disciplinary, systems-based approaches; its spanning the range of spatial scales from global to local; and its commitment to the "coproduction" of knowledge by researchers and practitioners (Clark and Dickson, 2003; Kauffman, 2009).

The systems approach is both formidable and necessary, in science as in policy making. Human–environment systems are complex, nonlinear, heterogeneous, spatially nested, and hierarchically structured (Wu and David, 2002). Feedback loops operate, multiple stable states typically exist, and surprises are inevitable (Kates and Clark, 1996). Change has multiple causes, can follow multiple pathways leading to multiple outcomes (Levin, 1998), and depends on historical context (Allen and Sanglier, 1979; McDonnell and Pickett, 1990). One important attribute of systems is their resilience, the system's ability to maintain structure and function in the face of perturbation and change. A second key attribute is the system's level of vulnerability: its exposure to hazards (perturbations and stresses) and its sensitivity and resilience when experiencing such hazards (Turner et al., 2003).

The systems approach to science is ideally suited to supporting sustainable management, both in advancing fundamental scientific understanding and in informing real-world decisions. It underlines the importance of linkages among various players at different scales, such as government agencies, private firms, citizen groups, and others.

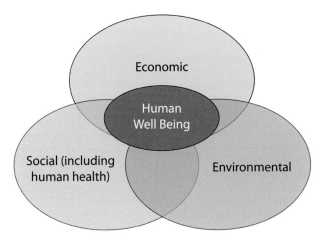

FIGURE 1-1 The components or domains of sustainability that support human well-being. SOURCE: National Research Council, 2011. Adapted from Figure 3-3, Hecht, 2010.

A healthy natural environment, though not the only component of sustainability, is an essential one; clean air, abundant and clean fresh water, biodiversity of plants, fish, and wildlife, and robust, highly-functioning ecosystems are all desired aspects of a healthy environment. In addition to maintaining a healthy environment, a sustainable society also provides systems to support other important societal values, including strong systems for preventive care and health care, public safety, transportation, energy, education, and housing. Societies also need strong economies in order to flourish.

All of these components interact with and depend upon one another. Social cohesion and effective legal systems are needed for economies to function efficiently; for example, a healthy and robust social fabric helps to ensure the health and well-being of people. Economic and social systems all interact with the environment, through natural resource services and extraction, food production, water systems, and natural biodiversity.

An approach to sustainability that includes human well-being provides a unifying framework for evaluating sustainability efforts. Moreover, this approach has intuitive appeal to policy makers and members of the public, who value human well-being in assessing environmental, economic, and social trade-offs.

Sustainability creates greater value, minimizes unintended consequences and ultimately improves the efficiency of government activities (see Box 1-2 and Box 1-3 for examples of federal agencies whose sustainability efforts have resulted in efficient use of resources and cost savings). Promoting sustainability reduces costs over the long term, which supports the economy and quality of life. The private sector has also embraced sustainability as a cost-effective organizing principle (see Box 1-4).

BOX 1-2
Sustainability at National Aeronautics and
Space Administration Facilities

National Aeronautics and Space Administration's (NASA's) sustainability policy is to execute the agency's mission "without compromising our planet's resources so that future generations can meet their needs. Sustainability involves taking action now to enable a future where the environment and living conditions are protected and enhanced. In implementing sustainability practices, NASA manages risks to mission, risks to the environment, and risks to our communities, all optimized within existing resources" (NASA, 2012). Some select sustainability objectives include: increasing energy efficiency and the use of renewable energy; measuring, reporting, and reducing direct and indirect greenhouse gas emissions; conserving and protecting water resources; eliminating waste, preventing pollution, and increasing recycling; and designing, constructing, maintaining, and operating high-performance sustainable buildings, among others (NASA, 2012).

Regarding the objective to design, construct, and maintain sustainable buildings, the Kennedy Space Center (KSC) has undertaken several such initiatives for its facilities, including those related to solar energy, waste diversion, and environmental remediation, which have resulted in efficient use of resources and significant cost savings. For example, KSC leased land to Florida Power & Light (FPL) in 2008 to build a 10-megawatt photovoltaic (PV) system for electricity generation. For use of the land, FPL provided KSC with a 1-megawatt PV system. This was cited as an innovative partnership that "helped the federal government and FPL electricity consumers achieve the environmental benefits of using electricity generated from renewable sources, and also helped NASA reduce energy costs that consume mission resources." With these innovations, the KSC facility is estimated to produce almost 1,800 megawatt-hours annually, saving the agency $162,221 in 2010. FPL's facility will produce nearly 19,000 megawatt-hours. The two systems will produce more than 560,000 megawatt-hours of electricity, saving KSC about $10.7 million during its expected 30-year life (NASA, 2011).

KSC achieved a solid waste diversion rate of 56.21 percent in 2010 by recycling and reusing construction and office material, which has saved the agency money. For example, the Coastal Revetment Project at KSC used recycled materials to replace an old decaying system with a new sustainable one. The 2.2-mile project incorporated 23,000 tons of concrete originating from demolished facilities, which saved about $3 million in project material costs. Additionally, the Environmental Remediation Program at the KSC embraced elements of sustainable green remediation into projects, primarily through the alternative power and bioremediation. For example, the agency successfully decontaminated groundwater at nine Kennedy sites. "At the GSA Seized Property Yard, bioremediation saved an estimated $400,000 compared to a traditional pump-and-treat system" (NASA, 2011).

RESOURCE CONNECTIONS AND GOVERNANCE LINKAGES

Concerns about Earth's sustainability in a form desirable to human habitation and quality of life traditionally rest on potential constraints to individual

---

BOX 1-3
Sustainability and the Department of Defense

The mission of the Department of Defense (DOD) is "to provide the military forces needed to deter war and protect the security of our country" (DOD, 2011). To successfully execute this mission, the military must have access to the energy, land, air, and water resources necessary to train and operate. According to DOD, "sustainability provides the framework necessary to ensure the longevity of these resources, by attending to energy, environmental, safety, and occupational health considerations" (DOD, 2011). Incorporating sustainability into DOD planning and decision making enables the agency to address current and emerging mission needs.

Within DOD, the Department of Army is responsible for achieving sustainability goals, including those related to renewable energy, in a fiscally prudent manner. The Army also serves as a test bed for developing and introducing new technologies for addressing sustainability challenges (Kidd, 2011). For example, the Army is leveraging available private-sector investment, including using power purchase agreements; enhanced-use leases; energy savings performance contracts; and utilities energy service contracts as tools to meet its objectives (Department of the Army, 2010). Regarding sustainable energy initiatives, the Army is pursuing initiatives such as utilizing waste energy or re-purposed energy using exhaust from boiler stack, building, or other thermal energy (Department of the Army, 2010).

In addition, to support renewable energy goals, the secretary of the Army established the Energy Initiatives Task Force (EITF) on August 10, 2011, with the mission to "identify, prioritize and support the development and implementation of large-scale, renewable and alternative energy projects"—focusing on attracting private investments and delivering the best value to the Army enterprise (Kidd, 2011). EITF serves as the central managing office for the development of large-scale Army renewable energy projects.

EITF is part of the Assistant Secretary of the Army for Installations, Energy and Environment (ASAIEE) that establishes "policy, provides strategic direction and supervises all matters pertaining to infrastructure, Army installations and contingency bases, energy, and environmental programs to enable global Army Operations" (ASAIEE, 2012). In order to respond to federal laws and energy directives/strategies of DOD, the Army needs to coordinate energy goals with environmental and sustainability goals. "An enterprise-wide approach is necessary because cost-effective management of energy requires coordinated efforts across the Army" and the optimization of limited resources to ensure success (Army Senior Energy Council, 2009).

---

natural resources. Rising prices resulting from resource scarcities generally have been shown to motivate technological innovations and substitutions that constrain the likelihood of 'running out' of resources (Krautkraemer, 2005). However, the continued presence of externalities associated with the extraction and use of natural resources suggests that their management to achieve a blend of economic, environmental, and socially sustainable outcomes will not result solely

---

BOX 1-4
BASF: Integrating Sustainability into Business Practices
A Private Sector Example

BASF, a global chemical company, has embraced sustainability as an organizing principle, stating that it has "strategically embedded sustainability" into the company as "a significant driver for growth" (BASF, 2013a). BASF defines sustainability as "balancing economic success with social and environmental responsibility, both today and in the future" (BASF, 2012; BASF, 2013a). The company has integrated sustainability into its core processes, including into the development and implementation of business units' strategies and research projects. It has also incorporated sustainability criteria into auditing processes for investment decisions (BASF, 2012).

Sustainability issues are identified by the company using material analysis; top priority issues include energy and climate, water, renewable resources, product stewardship, human capital development, human and labor rights, and biodiversity (BASF, 2012). The company states that sustainability management involves "taking advantage of business opportunities, minimizing risks and establishing strong relationships with our stakeholders" (BASF, 2012).

As a result, BASF reported that in 2012, the company reduced its greenhouse gas emissions by 31.7 percent per metric ton of sales product and increased its energy efficiency by 19.3 percent compared with baseline 2002. Similarly, in 2012, the total emissions of air pollutants from the chemical plants into the atmosphere dropped by 63.1 percent to 31.580 metric tons (BASF, 2013b).

---

from commodity price signals (Krautkraemer, 2005; Tietenberg, 2005). It is obvious that these constraints are real and, in many cases, problematic. Here are several examples:

- Constraints on traditional energy supplies[1] and challenges related to climate will require a transition to a broader mix of fuels over the next several decades, consistent with reducing greenhouse gas emissions and other environmental impacts (NRC, 2009; Chu and Majumdar, 2012). While market signals drive innovations in energy technologies and can influence the search for energy substitutes, the continued presence of externalities and impacts on environmental goods such as biodiversity, air quality, and so on, associated with energy generation and use suggest the need for a decision framework and policies that incorporate and integrate these multiple considerations. Major efforts will be required because the required changes are so huge.

- Global demand for nonrenewable resources such as metals is rising rapidly, mainly in developing economies. Concomitantly, the use of progressively poorer ore grades will become a real problem in the future as demand and pro-

---

[1]For example, there are geographical, geological, economic, legal, and environmental constraints on the future use of coal. The National Research Council's report *America's Energy Future* provides excellent reviews of these topics.

duction increase, requiring ever more energy and water to enable ore processing (MacLean et al., 2010). The rise in demand for certain rare earths also implies that for the foreseeable future, recycling will not provide an important supplementary resource for these minerals. While increasing scarcities will likely drive up prices and stimulate development of substitutions, accessing traditional, poorer grade, or new metals all involve impacts to lands, potentially impact wildlife, and can affect other environmental amenities.

• Population growth and improving quality of life are expected to place increased pressure on productive land, risking the loss of important ecosystems and their beneficial functions.

• In addition, substantial growth is expected in global freshwater use. Consequently, the quality and quantity of available freshwater per capita will decrease in certain localities in the absence of significant changes in water management and use patterns.

Other constraints deserve consideration, especially those resulting from limitations involving connections among the resources. Although resource sustainability is a problem generally approached in a piecemeal fashion, it is a systems problem, and the links that connect the resources are often more challenging to address than those of the individual resources themselves. It may help to picture the challenge of sustainability as shown in Figure 1-2, where key resource domains, including water, land, energy and non-renewable resources, are shown as squares, and areas that require these resources (industry, agriculture, nature, and domestic) are depicted as ovals. Human health and well-being interacts with all of these. It is common that scientists and decision makers specialize in one of these topics and are relatively unaware of the important constraints that may occur as a result of inherent connections with other topics. A near-complete linkage exists among all of these areas, yet tradition and specialization encourage a focus on a selected oval and all of the squares or to a selected square and all of the ovals (Graedel and van der Voet, 2010). Graedel and van der Voet (2010) pose the question: Can we devise an approach that addresses them all as a system, to provide the basis for constructing a coherent package of actions that optimize the system, not the system's parts?

CONNECTIONS: THE SCIENTIFIC CHALLENGE OF UNDERSTANDING SYSTEMS

In modern society, the interrelatedness of the natural and human worlds is even more complex. The systems that must be considered in addressing sustainability challenges are referred to in this report as social-ecological systems.[2] These complex systems include the natural resource domains (air, fresh water,

---

[2]The term social-ecological systems is an increasingly used research framework. Ostrom E. 2009. A general framework for analyzing sustainability of social-ecological systems. *Science* 325(5939):419-422.

coastal oceans, land, forests, soil, etc), built environments (urban infrastructure such as drinking water and wastewater systems, transportation systems, energy systems), and the social aspects of complex human systems (such as public health, economic prosperity, and the like).

These elements of social-ecological systems are all interconnected, and the sustainability challenges that the nation faces rarely involve only one of them. Furthermore, the impacts of indirect connections to supply chains for manufactured and agricultural goods, or the connection to externalities such as the costs of the loss of ecosystem services, might also need to be factored in when addressing sustainability challenges.

Some connections are obvious. A coal-fired power plant provides electricity, which provides social, economic, and health benefits, but it also expends a nonrenewable resource, uses water to provide steam, emits products of combustion into the air, and generates solid waste. Some connections are less obvious. Battery-powered vehicles have no direct emissions to the atmosphere at the time of use, an apparent advantage over internal combustion vehicles. However, generating the electricity to charge the battery has impacts that may occur far away from where the vehicle is used. Also, disposal of battery can increase emissions due to energy consumed in recovering and recycling the materials.

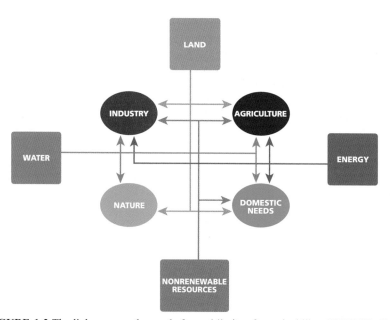

FIGURE 1-2 The links among the needs for and limits of sustainability. SOURCE: Graedel, T. E., and E. van der Voet, 2010, adapted from Figure 1.2 The links among the needs for and limits of sustainability. Reprinted with permission from the MIT Press.

Some connections become apparent only over time. Use of persistent pesticides in the production of crops in the 1950s was effective; however, some of the pesticides were eventually found to persist, bioaccumulate, and have long-term effects on higher species only after some period of use. Similarly, studies have indicated that exposure to endocrine disruptors during critical periods of development can cause delayed effects that do not become evident until later in life (European Commission, 2011). We call these types of situations temporal connections.

Connections that are indirect can nonetheless be highly significant. Demand for ethanol in the United States caused the price of corn to rise and caused a shift in land use from soybean production to corn production. To fill the void, land was deforested in other countries and planted in soybeans. This is an example of a spatial connection. Other connections occur when multiple demands for the same resources are influenced through economic markets.

Consider the example of the sustainability challenge of growing sufficient food while also developing renewable energy from biofuels—the so-called food vs. fuel debate. The connections that must be considered include (1) the amount and type of land used to grow crops for food and that used to grow biofuels; (2) water use for crops as well as for biofuel production, transportation infrastructure use and costs for transporting both; (3) the relative impact of greenhouse gas emissions (including the emissions from indirect aspects of the system, such as emissions associated with growing and transporting the crops and producing the food and biofuels, as well as emissions from end use of the crops and fuels); (4) impacts on energy consumption to produce the food and fuel (again including indirect aspects); (5) the impacts on food cost and its availability to all economic classes of the U.S. public; (6) the impact on local economies as well as the export and import of food and fuel; and (7) limited time offer government subsidies and longer term sustainable farming practices, such as crop rotation.

The examples that the committee studied all reflected the interconnections among social-ecological systems. In Philadelphia, for instance, the effort to manage stormwater more sustainably by investing in green infrastructure[3] rather than storm sewers is not just a water issue; it has impacts on air quality (through green plantings), energy consumption (water infrastructure), community well-being (through the creation of rain gardens), and neighborhood violence (through the greening of abandoned and overgrown lots). These connections are explained in more detail in Chapter 3.

The Mojave Desert, discussed as another example, is used for recreation, housing, and military training and is a premium location for renewable energy development, as it has some of the highest-quality solar and wind resources in the nation. It is also home to mining, agriculture, and human communities, as well as unique ecosystems and a number of endangered species. The competition between human-centric land uses and the desire to preserve species habitat

---

[3]Green infrastructure refers to the management of stormwater runoff through the use of natural systems.

and manage on an ecosystemwide basis has increased the need for coordinated land management in the Mojave Desert. The interconnections in the Mojave Desert example were evident in conflicts over competing land uses. One cannot successfully address sustainability issues in a specific social-ecological system without first identifying the relevant connections.

LINKAGES: THE GOVERNANCE CHALLENGE
OF MANAGING CONNECTED SYSTEMS

While addressing connections across natural and human system domains may be challenging, successful governance requires it. Ignoring connections raises the risk of policy actions that result in unintended consequences and ineffective and inefficient outcomes. For example, pursuit of policies to augment use of lands for biofuels production will have impacts on water use, food production, and wildlife. Unless these connections are assessed, policies and investments to promote biofuels could have unintended impacts on food commodity prices and water availability (Tilman et al., 2009). On the other hand, sustainability approaches that optimize a bundle of benefits could help meet energy needs while simultaneously reducing greenhouse gas emissions, sustaining biodiversity, and enhancing food security. Sustainable management of connected systems calls for governance that effectively links across domains, as well as across geographic and temporal scales.

The strong organizational linkages needed to support sustainability approaches can be extraordinarily difficult to implement. Political realities sometimes run counter to scientific and technical currents. As political scientist Eugene Bardach (1998) wrote, "Political and institutional pressures on public sector agencies in general push for differentiation rather than integration, and the basis for differentiation is typically political rather than technical." These challenges are the subject of Chapter 2, and possible solutions are examined in Chapter 5.

REFERENCES

Allen, P. M., and M. Sanglier. 1979. Dynamic-model of growth in a central place system. *Geographical Analysis* 11:256.

Army Senior Energy Council and the Office of the Deputy Assistant Secretary of the Army for Energy and Partnerships. 2009. Army Energy Security Implementation Strategy. Washington, DC: U.S. Army.

Assistant Secretary of the Army for Installations, Energy and Environment. 2012. Online. Available at http://www.army.mil/ASAIEE. Accessed February 28, 2013.

Bardach, E. 1998. Getting Agencies to Work Together: The Practice and Theory of Managerial Craftsmanship. Washington, DC: Brookings Institution.

BASF. 2012. BASF Report: Economic, Environmental, and Social Performance. Online. Available at http://www.basf.com/group/corporate/en/function/conversions:/publish

download/content/about-basf/facts-reports/reports/2012/BASF_Report_2012.pdf. Accessed March 11, 2013.

BASF. 2013a. Sustainability. Online. Available at http://report.basf.com/2012/en/manage mentsanalysis/sustainability.html. Accessed March 11, 2013.

BASF. 2013b. BASF with positive results in goals for environment, health and safety. Online. Available at http://www.basf.com/group/pressrelease/P-13-157. Accessed March 11, 2013.

CERES (Coalition of Environmentally Responsible Economies). 1989. The Ceres Principles. Online. Available at http://www.ceres.org/about-us/our-history/ceres-principles. Accessed October 1, 2012.

Chu, S. and A. Majumdar. 2012. Opportunities and challenges for a sustainable energy future. *Nature* 488:294-303.

Clark, W. C., and N. M. Dickson. 2003. Sustainability science: The emerging research program. *Proceedings of the National Academy of Sciences USA* 100(14):8059-8061.

Department of the Army. 2010. Army Vision for Net Zero. Office of the Assistant Secretary for the Army. February 17, 2010.

Department of the Army. 2012. Energy Goal Attainment Responsibility Policy for Installations. Online. Available at http://www.asaie.army.mil/Public/Partnerships/EnergySec urity/docs/ASAIEE_energy_goal_attainment_policy_24_Aug_2012.pdf. Accessed February 28, 2013.

DOD (U.S. Department of Defense). 2011. Strategic Sustainability Performance Plan. Online. Available at http://www.denix.osd.mil/sustainability/upload/dod-sspp-fy1 1-final_oct11.pdf. Accessed March 26, 2013.

Ecologically Sustainable Development Steering Committee Endorsed by the Council of Australian Governments. 1992. National Strategy for Ecologically Sustainable Development. Online. Available at http://www.environment.gov.au/about/esd/publica tions/strategy/intro.html. Accessed September 28, 2012.

Environment Canada. 2010. Planning for a Sustainable Future: A Federal Sustainable Development Strategy for Canada, Consultation Paper. Gatineau, Quebec: Federal Sustainable Development Office.

EC (European Commission). 2011. State of the Art Assessment of Endocrine Disruptors. Final Report. Project Contract Number 070307/2009. Online. Available at http:// ec.europa.eu/environment/endocrine/documents/4_SOTA%20EDC%20Final%20R eport%20V3%206%20Feb%2012.pdf. Accessed February 19, 2013.

Fiksel, J. 2006. Sustainability and resilience: Toward a systems approach. *Sustainability: Science, Practice, and Policy* 2(2).

Graedel, T. E., and E. van der Voet. 2010. Linkages of sustainability: An introduction. Pp. 1-10 in Linkages of Sustainability, T. E. Graedel and E. van der Voet, eds. Cambridge, MA: MIT Press.

Gunderson, L., and C. S. Holling. 2002. Panarchy: Understanding Transformations in Human and Natural Systems. Washington, DC: Island Press.

ICLEI—Local Governments for Sustainability USA. 2010. STAR Community Index: Sustainability Goals and Guiding Principles. Online. Available at http://www.iclei usa.org/library/documents/STAR_Sustainability_Goals.pdf. Accessed October 1, 2012.

Kates, R., and W. Clark. 1996. Expecting the unexpected? *Environment* 38:6.

Kates, R., W. Clark, R. Corell, J. Hall, C. Jaeger, I. Lowe, J. McCarthy, H-J. Schellnhuber, B. Bolin, N. Dickson, S. Faucheux, G. Gallopin, A. Grubler, B. Huntley, J. Jager, N. Jodha, R. Kasperson, A. Mabogunje, P. Matson, and H. Mooney. 2001. Sustainability science. *Science* 292(5517):641-642.

Kauffmann, J. 2009. Advancing sustainability science: report on the International Conference on Sustainability Science (ICSS) 2009. *Sustainability Science* 4(2):233-242.

Kidd, R. G. IV. 2011. Department of Defense Perspective on Sustainability Linkages. Presentation to the National Research Council's Committee on Sustainability Linkages in the Federal Government, First Meeting. September 20, 2011.

Krautkraemer, J. A. 2005. Economics of Resource Scarcity: The State of the Debate. Discussion Paper, April 2005. Washington, DC: Resources for the Future.

Leggett, J. A., and N. T. Carter. 2012. Rio+20: The United Nations Conference on Sustainable Development, June 2012. *Congressional Research Service* 7-5700.

Levin, S. A. 1998. Ecosystems and the biosphere as complex adaptive systems. *Ecosystems* 1:431-436.

Lewis, C. S. 1947. Miracles: A Preliminary Study. 1st Ed. London: Geoffrey Bles.

McDonnell, M. J., and S. T. A. Pickett. 1990. The study of ecosystem structure and function along urban-rural gradients: an unexploited opportunity for ecology. *Ecology* 71:1231-1237.

MacLean, H. L., F. Duchin, C. Hagelüken, K. Halada, S. E. Kesler, Y. Moriguchi, D. Mueller, T. E. Norgate, M. A. Reuter, and E. van der Voet. 2010. Stocks, Flows, and Prospects of Mineral Resources. Pp. 199-218 in Linkages of Sustainability. T. E. Graedel and E. van der Voet, eds. Cambridge, MA: MIT Press.

Meadows, D. H., D. L. Meadows, and J. Randers. 1992. Beyond the Limits. White River Junction, VT: Chelsea Green Publishing.

NASA (National Aeronautics and Space Administration). 2011. Kennedy Space Center's Sustainability Initiatives. Online. Available at http://www.nasa.gov/centers/kennedy/pdf/566523main_sustainability-initiatives.pdf. Accessed February 28, 2013.

NASA. 2012. Strategic Sustainability Performance Plan. Online. Available at http://www.nasa.gov/pdf/724131main_NASA_SSPP%202012%20abridged.pdf. Accessed February 28, 2013.

NEPA (National Environmental Protection Act of 1969). 2000. Online. Available at http://epw.senate.gov/nepa69.pdf. Accessed September 28, 2012.

NRC (National Research Council). 2009. America's Energy Future: Technology and Transformation. Washington, DC: National Academies Press.

NRC. 2011. Sustainability and the U.S. EPA. Washington, DC: National Academies Press.

NRC. 2012. Ecosystem Services: Charting a Path to Sustainability. Washington, DC: National Academies Press.

OECD (Organisation for Economic Co-operation and Development). 2007. OECD Sustainable Development Studies: Institutionalising Sustainable Development. Paris, France: OECD.

OECD. 2009. Declaration on Green Growth (Adopted at the Council Meeting at Ministerial level on June 25, 2009). Online. Available at http://search.oecd.org/officialdocuments/displaydocumentpdf/?doclanguage=en&cote=C/MIN(2009)5/ADD1/FINAL. Accessed August 30, 2012.

OECD. 2011. Towards Green Growth: Green Growth Strategy Synthesis Report. Online. Available at http://www.oecd.org/officialdocuments/publicdisplaydocumentpdf/?cote=C/MIN(2011)4&docLanguage=En. Accessed August 30, 2012.

Ostrom, E. 2009. A general framework for analyzing sustainability of social-ecological systems. *Science* 325(5939):419-422.

Parliamentary Office of Science and Technology. 2012. Seeking Sustainability. *POSTnote* 408.

PCAST (President's Council of Advisors on Science and Technology). 2011. Sustaining Environmental Capital: Protecting Society and the Economy. Washington, DC: Executive Office of the President.

Southern Growth Policies Board. Landfill Gas Project. Online. Available at http://www.southernideabank.org/items.php?id=2601. Accessed October 30, 2012.

Skaggs, R., K. Hibbard, P. Frumhoff, T. Lowry, R. Middleton, R. Pate, V. Tidwell, J. Arnold, K. Averyt, A. Janetos, C. Izaurralde, J. Rice, and S. Rose. 2012. Climate and Energy-Water-Land System Interactions: Technical Report to the U.S. Department of Energy in Support of the National Climate Assessment. Richland, WA: Pacific Northwest National Laboratory.

The White House. 2000. Executive Order 13148 of April 21, 2000. Greening the Government Through Leadership in Environmental Management. *Federal Register* 65(81):24595-24606.

The White House. 2007. Executive Order 13423 of January 24, 2007. Strengthening Federal Environmental, Energy, and Transportation Management. *Federal Register* 72(17):3919-3923.

The White House. 2009. Executive Order 13514 of October 5, 2009. Federal Leadership in Environmental, Energy, and Economic Performance. *Federal Register* 74(194): 52117-52127.

Tietenberg, T. 2005. Environmental and Natural Resources Economics. 7th ed. Boston, MA: Addison Wesley Longman.

Tilman, D., R. Socolow, J. A. Foley, J. Hill, E. Larson, L. Lynd, S. Pacala, J. Reilly, T. Searchinger, C. Somerville, R. Williams. 2009. Beneficial Biofuels—The Food, Energy, and Environment Dilemma. *Science* 325:270-271.

Turner, B. L., R. E. Kasperson, P. A. Matson, J. J. McCarthy, R. W. Corell, L. Christensen, N. Eckley, J. X. Kasperson, A. Luers, M. L. Martello, C. Polsky, A. Pulsipher, and A. Schiller. 2003. A framework for vulnerability analysis in sustainability science. *Proceedings of the National Academy of Sciences USA* 100:8074-8079.

UK Department for Environment Food and Rural Affairs. 2005. Securing the future: delivering UK sustainable development strategy. Online. Available at http://www.defra.gov.uk/publications/files/pb10589-securing-the-future-050307.pdf. Accessed October 1, 2012.

UNEP (United Nations Environment Programme). 2002. Melbourne Principles for Sustainable Cities. Online. Available at http://www.iclei.org/fileadmin/user_upload/documents/ANZ/WhatWeDo/TBL/Melbourne_Principles.pdf. Accessed October 1, 2012.

World Commission on Environment and Development. 1987. Our Common Future (The Brundtland Report). Oxford: Oxford University Press.

Wu, J., and J. L. David. 2002. A spatially explicit hierarchical approach to modeling complex ecological systems: theory and applications. *Ecological Modeling* 153:7-26.

Chapter 2

The Impediments to Successful Government Linkages

As discussed in Chapter 1, issues of sustainability are by their nature complex, involving multiple domains, multiple exigencies, multiple locations, and multiple time frames. But the federal government is generally not organized or operated to deal with this complexity. Rather, for a variety of reasons, federal agencies generally focus on one arena (e.g., health, energy, environment), with programs addressing one exigency (e.g., natural disaster, statute) in one domain (e.g., air, water, land use) and one time frame (e.g., dictated in statute, term of office). Specifically, impediments or barriers to successful governance linkages come from legal limitations in the form of structural or vertical fragmentation of authority; funding mechanisms that favor short-term, single-agency initiatives rather than longer-term cross-agency projects; a lack of access to or coordination of such foundational elements as research and information/data; and the culture of government. The absence of a national sustainability policy means there are very few institutional bridges, practices, or processes that incentivize building and sustaining the necessary linkages. The difficulties of creating or forging such ties were evident in many of our fact-finding examples, as were the ways in which such impediments or barriers could be overcome. The examples provide a basis for our recommendations for improving the efficiency and effectiveness of governance linkages discussed in Chapter 5.

AUTHORIZATIONS—FRAGMENTED AND DIFFUSE

One of the most significant challenges to governance linkages is that the basic framework of government, established by law, is one of separated and dispersed authority, and an essential part of that framework is that government agencies at all levels—federal, state, local, tribal and even international—can only do what they have been authorized to do by their governing authorities—namely, Congress, state legislatures, etc. (U.S. Supreme Court, 1988). Some authorizing statutes provide general grants of authority, but others are quite prescriptive, limiting the agency to the terms of the statute (for example, the 1976 Resource Conservation and Recovery Act).

In addition, many authorizing or enabling acts focus on a single mission or a single domain—water or energy, for example—even if the domain is part of an interconnected resource system. For example, as noted in Chapter 1, water and energy are interrelated in many circumstances, yet water quality is within the jurisdiction of the U.S. Environmental Protection Agency (EPA), and water supply and flow are largely governed by state law and several federal agencies, including the U.S. Army Corps of Engineers and the Bureau of Reclamation (BOR) at the Department of the Interior (DOI), while energy is within the jurisdiction of the Department of Energy (DOE), the Bureau of Land Management (BLM), and the Bureau of Ocean Energy Management at DOI, as well as the Federal Energy Regulatory Commission. Nonetheless, within this complex institutional context and statutory provisions, there is often delegated discretion and hence maneuverability—so-called "white space"—for creative or innovative individuals and agencies.

The pejorative, but nonetheless accurate, description for this fragmentation of authority is the stovepipe or silo effect: Each agency focuses on implementing its own statutory mandate (Kettl, 2002). There is a rational justification for this phenomenon—namely, the agencies were created as repositories for expertise and experience in a particular area (Kagan, 2001), so they appropriately concentrate on that area and not on matters outside their jurisdiction. But there are consequences for concentrating in this way: It often leads to silo-based approaches to interconnected systems.

Within each agency, there are scientists, economists, engineers, lawyers, and other personnel all focused on the same set of issues, enabling the agency to bring an interdisciplinary approach to problem solving in that domain. Their reach does not, however, extend to connected domains. In fact, when two or more agencies (or distinct parts of a single agency[1]) share responsibility for an interconnected system or a single geographic location, the reality of fragmented authority is a huge challenge for successful collaboration. Multiple agencies often share responsibility for a domain. Consider the Mojave Desert, where the land is subject to the U.S. Department of Agriculture's (USDA) Forest Service; four separate agencies within DOI, including BLM, Bureau of Indian Affairs, U.S. Fish and Wildlife Service (FWS), and the National Park Service (NPS); DOE (which has responsibility for a large alternative energy facility); and the Department of Defense (DOD) (which has six military bases with all branches of the service present). The presence of multiple agencies means multiple and sometimes conflicting goals, as well as multiple leaders who have to sort out their respective roles and responsibilities while remaining faithful to their agencies' mandates.

---

[1]For example, both air and water are within the jurisdiction of EPA, although each is housed in a separate office, operating under two separate statutes, the Clean Air Act, 42 U.S.C. Sec. 7401 et. seq, and the Clean Water Act, 33 U.S.C. Sec 1251 et.seq.

There may also be challenges even when a single agency has sole place-based authority, for example, land management agencies such as BLM or the Forest Service, which have responsibility for air, water, species and habitat protection, mineral resources, grazing and energy leasing, recreational uses, and other activities for a particular geographic area. These cases can be challenging because the issues the agencies must deal with, particularly if they are undertaking activity on that land, are subject to statutes (often prescribing one-size-fits-all approaches) enforced by a sister agency.[2] Executive Branch policy has been made clear through Executive Orders in effect for several decades[3] that *all* activities on federal land are subject to *all* applicable federal and state laws and regulations.[4] In these cases, agencies may be limited to activities prescribed in statutes, such as one-size-fits-all approaches that may not be in line with sustainability efforts. In addition, the place-based agency often has to interact with its sister agencies that have different agendas and priorities, which can lead to conflict rather than collaboration.

Federal agencies are subject not only to substantive laws, but also to procedural laws that affect the way they may address issues. One of the most significant is the Administrative Procedure Act, 5 U.S.C. Sec. 551 et seq., which sets forth requirements for issuing regulations. The most prevalent regulatory process is "informal" rulemaking under 5 U.S.C. Sec. 553, a process that has the advantage of promoting public participation—an element that is critical to successfully addressing sustainability connections, but which is also very time consuming and costly. The Federal Advisory Committee Act, 5 U.S.C. App. 2, also prescribes how federal agencies can work with stakeholders and advisors from outside the government; regrettably for purposes of enhancing engagement of affected entities, the Act has in practice often proved to be confining and rigid, even when the need for inclusiveness and flexibility are at a premium.[5] Another type of constraint comes from federal acquisition laws, which generally require competitive bidding for the acquisition of goods and services.[6] The provisions

---

[2]For example, the Clean Air Act, 42 U.S.C. Sec. 7401 et seq., and the Clean Water Act, 33 U.S.C. Sec. 1251, et seq., enforced by the EPA or the Endangered Species Act, 16 U.S.C. Sec. 1531 et seq., enforced by FWS and the National Oceanic and Atmospheric Administration (NOAA).

[3]Executive Order (EO) 13148, Greening the Government Through Leadership in Environmental Management (2000) and EO 13514, Federal Leadership in Environmental, Energy and Economic Performance (2009).

[4]One notable exception is for border security infrastructure, in which the Secretary of the Department of Homeland Security was given special statutory authority to waive all other laws in implementing the provisions of the act.

[5]Among other things, while the Act provides greater transparency, it often operates to restrict the number and role of involved stakeholders and may impose significant costs and time delays on the sponsoring agency.

[6]See Federal Acquisition Regulation (FAR) 2.101, 35.003…266 and 41 U.S.C.2302-2213 (1994) and Federal Grant and Cooperative Agreement Act of 1979 (31 U.S.C.6301 et seq.) For a discussion of use of cooperative agreements and challenges in their use, see

are wholly appropriate to ensure best value for federal dollars when purchasing big ticket items, but may not be well suited to implementing multiparty, multi-issue consensus-based partnership activities, where an expenditure of funds is an integral ingredient.[7]

In addition to the horizontal structural fragmentation that comes from federal agencies operating as separate and distinct silos, there are also vertical structural challenges, especially when statutes set forth explicit roles both for the federal government and for affected state, local, and/or tribal governments. For example, our environmental laws require EPA to set standards that are implemented or enforced by state, local, and tribal governments.[8] This kind of shared responsibility is the norm, not the exception, throughout the regulatory world, from agriculture to transportation and from health to housing.

Shared responsibility is also the norm in place-based contexts. A central driver of many landscape-scale sustainability challenges is land use. Yet land-use decision authority is typically widely distributed among many federal agencies on public lands, is shared with state land authorities, and is highly distributed among local and private interests. Drawing again on the Mojave Desert example, there were, in addition to the interests of the federal agencies listed above, 10 state parks, eight county jurisdictions, and 37 federally recognized Native American Indian tribes on the land.[9]

It is also important to recognize that governmental entities do not and cannot operate in a vacuum. Rather, in any project involving environmental, economic, and social interconnections, there will also invariably be a number of nongovernmental organizations (NGOs) as well as private-sector entities who will have a stake in the outcome and whose decisions and actions vitally affect the success of federal agency activities. The participation of these entities can be

---

DOI Office of Inspector General. 2007. Proper Use of Cooperative Agreements Could Improve Interior's Initiatives for Collaborative Partnerships. Report No. W-IN-MOA-0086-2004. For implementing policies, see OMB. 1978. Implementation of the Federal Grant and Cooperative Agreement Act of 1977. Federal Register 43 (161):36860 and OMB Circulars.

[7]In these cases, in providing funding to a partner, the federal government is not purchasing a good or service but rather working with the partner to provide a public good. There are tools such as cooperative agreements available for use in these partnerships, but the circumstances in which such agreements can be used is both unclear and limited in some circumstances.

[8]For example, see U.S. Court of Appeals for the District of Columbia Circuit. 2012. EME Homer City, L.P. v EPA et al., No 11-1302.

[9]While not the focus of this report, international institutions and other national governments are sometimes involved in domestic sustainability connections. Consider the Great Lakes example involving water matters as well as birds that migrate across international boundaries. The Great Lakes project itself is the subject of an international treaty, and representatives of the Canadian government at various levels work side by side with U.S. federal and state officials.

critical to progress. NGO participation in the committee's examples ranged from small local organizations such as the Philadelphia Horticultural Society to larger national organizations such as the Nature Conservancy. The USDA's Sustainable Agriculture Research and Education (SARE) program, a decentralized competitive grants and education program operating in every state and island protectorate, is one of the federal programs that has engaged stakeholders substantively both in technical review process and within the grants themselves (mandatory) since 1988 (SARE, 2013).

As suggested above, one consequence of fragmented authority is that there is often overlapping or even conflicting authority for a particular initiative. At the other end of the spectrum is when the various entities do join together but the resulting "coordinating committee" or "managing directorate" is without any formal authority. This was the limitation faced by the Desert Managers Group (DMG) in the Mojave Desert—a situation that ultimately prompted California to create a legal entity, the Desert Renewable Energy Conservation Group, complete with regulatory authority.

FUNDING CHALLENGES

Government entities can undertake activities only if funds are appropriated for those activities. Yet budgets are prepared on an agency-by-agency basis, and agencies typically promote and defend their own initiatives rather than multiagency initiatives. There are a few existing vehicles available for cross-agency funding,[10] and occasionally agencies prepare cross-cut budgets for high-priority initiatives of an administration. However, such ventures are often difficult to administer. Even when the requests for funding multiagency projects are accompanied, as they generally are, with a solid business case that demonstrates significant savings from a coordinated rather than unilateral agency approach (Kaufman, 2012), such requests are often resisted by budget overseers.

The role of budget overseers cannot be overstated. While budgets are prepared by the agencies, and after review and modification by the Office of Management and Budget (OMB) presented as the President's budget, the actual federal budget is what is ultimately enacted by the Congress and signed by the

---

[10]One successful and longstanding program (operating since 1978) is the National Toxicology Program. The program is housed at the National Institute of Environmental Health Sciences (NIEHS) but is supported by NIEHS, the National Institute for Occupational Safety and Health (NIOSH), and the Food and Drug Administration (FDA). The Executive Board has broader agency representation. The Executive Committee is made up of EPA, the Consumer Product Safety Commission (CPSC), DOD, National Cancer Institute (NCI), Centers for Disease Control and Prevention's (CDC's) National Center for Environmental Health/Agency for Toxic Substances and Disease Registry (NCEH/ATSDR), Occupational Safety and Health Administration (OSHA), NIOSH, NIEHS, and FDA. See http://ntp.niehs.nih.gov/?objectid=7201637B-BDB7-CEBA-F57E 39896A08F1BB. A less formal device is called "pass-the-hat," whereby several agencies are asked to contribute to a project on a per-use or per-capacity basis.

President. In practice, therefore, the fate of the President's budget is subject to the organizational and natural inclinations of the congressional appropriations committees. Some of the committees are congruent with one or more specific agencies,[11] while others have only a part of an agency.[12] Some committees have jurisdiction over connected domains[13] but not jurisdiction over all of the pieces or players in an interconnected resource system, including related economic or social aspects of the issue. Absent a national sustainability policy or a legal entity charged with developing or implementing such a policy, there are limited mechanisms to fund projects and programs designed to address sustainability issues.

In addition, congressional appropriations committees often seek to ensure that the funds they appropriate are spent by the agency within the committee's jurisdiction on the activities approved by that committee. NPS, for example, is generally prohibited from spending any of its funds outside of national parks unless otherwise specifically authorized at a particular park.[14] In the same way, committee members are reluctant to appropriate funds for matters they view as the responsibility of another committee, even if those matters relate to the mission of an agency that is within their jurisdiction. This line-of-sight approach enhances accountability because agencies get the message about the need to stay within their boundaries in the clearest, most concrete way, but it makes cross-agency funding appreciably more difficult, even for projects related to the agency's core mission.

The challenge for funding governance linkages is further exacerbated by the fact that budgets are approved on an annual basis, but most sustainability initiatives require efforts over many years' duration. While the private sector often embraces sustainability as an organizing principle because it can save

---

[11]For example, the Senate Appropriations Subcommittee on Energy and Water Development has jurisdiction over DOE programs, and various agencies that manage hydropower.

[12]For example, the U.S. Senate Appropriations Subcommittee on Interior, Environment and Related Agencies, oversees the budget for all Interior Department agencies except the BOR, which is under the jurisdiction of the Senate Appropriations Subcommittee on Energy and Water.

[13]Energy issues, which have a significant nexus with water, involve actions of the BLM, which comes under the jurisdiction of the Senate Appropriations Subcommittee on Interior, Environment, and Related Agencies, while the BOR, which has both energy and water responsibilities, falls under the jurisdiction of the Senate Subcommittee on Energy and Water.

[14]The Land and Water Conservation Fund limits acquisition spending to within park boundaries; operating funds for the park system have generally been interpreted in annual appropriations to be allowable only within the park system. Supporting this interpretation, parks seeking to invest outside of park boundaries on projects that benefit adjacent national parks have sought special authorizing legislation for those purposes.

money in the long term,[15] government appropriators are often driven by short-term results notwithstanding business cases predicting positive longer-term results. In the committee's experience, a further challenge is the implicit incentive budget managers feel to "use it or lose it" in order to maintain future funding. This approach to budget allocation leads to inefficient use of capital and can further complicate the ability to address sustainability over the long-term. In any event, the current fiscal environment is obviously putting enormous pressure on existing agency funding, and new money is very hard to come by, especially for discretionary programs that may not enjoy significant bipartisan support.

FRAGMENTATION OF FOUNDATIONAL ELEMENTS: INFORMATION AND RESEARCH

One of the observed consequences of fragmented authority and the silo effect is that agencies have traditionally generated or compiled the data they need or have undertaken research for activities they view as their own, independent of their sister agencies.[16] This approach presents additional impediments to creating and sustaining governance linkages.

First, even though data or research generated by an agency may be directly tied to that agency's particular needs and purposes, it could at the same time be invaluable to sister agencies and to the public. Indeed, in the Mojave Desert example, it appeared that not all offices within BLM, let alone the other federal agencies, have access to the same maps of the area for which they were responsible. Some agencies are sharing data in certain areas, and recent administration initiatives have further encouraged agencies to identify their useful data sets and, to the extent feasible, make them available online.[17] In specific situations, agen-

---

[15]For example, the Corporate Eco Forum (CEF) is a membership organization for large companies that have committed to the environment as a business strategy issue. The mission is to "help accelerate sustainable business innovation by creating the best neutral space for business leaders to strategize and exchange best-practice insights. Members represent 18 industries and have combined revenues exceeding $3 trillion" (CEF, 2012). A report released by the CEF in 2009 noted that research shows that "becoming environment-friendly lowers costs because companies end up reducing the inputs they use. In addition, the process generates additional revenues from better products or enables companies to create new businesses. In fact, because those are the goals of corporate innovation, we find that smart companies now treat sustainability as innovation's new frontier" (Nidumolu et al., 2009). A list of CEF member companies can be found at http://www.corporateecoforum.com/contact/index.php.

[16]For example, coordination among federal agencies in the critically important area of regional-scale water resource modeling could be greatly improved and expanded. Current efforts at NOAA and NWS would benefit from additional partnering with USGS, DOE, and other agencies that are pursuing individual efforts.

[17]For example, see Orszag, P. R. 2009. Memorandum for the Heads of Executive Departments and Agencies. Open Government Directive at: http://www.whitehouse.gov/open/documents/open-government-directive.

cies have also begun to create coordinated and integrated databases.[18] These are particularly valuable when several agencies share responsibility for a particular place or domain.

Nonetheless, combining (or having the ability to access) disparate data sets is not a panacea. Many agencies have specific statutory authorities that require the generation of specialized information that may not be readily combined with other data sets. Even when not set by statute, protocols for data collection and compilation vary widely across agencies, making data integration across domains and even within domains extremely challenging. For example, DOI, USDA and other agencies were interested in linking information on various aspects of water, but there were over a dozen significant databases housed in multiple agencies that contained both duplication and inconsistencies. As a result, agencies found it very difficult to merge these data and present coherent and cohesive information on water flow, quality, and other relevant variables.[19] In addition, any so-called commons for information presents its own problems. Agencies are often legitimately concerned about losing control of their data or having someone else assume responsibility for it. Moreover, there may be technical issues, privacy concerns, and the need to commit significant agency resources to post the data, ensure its accuracy, and maintain its currency. As a result, such common repositories are often underused. Cloud computing[20] has

---

[18] Examples include the Landscape Conservation Cooperative (which establishes 22 ecosystem regions through which federal agencies are working with states, tribes, local governments, NGOs, and the academic community to coordinate data, identify information gaps, and develop shared strategies for generating and using scientific information), with information available at: http://www.doi.gov/lcc/index.cfm; and LANDFIRE (which has attempted to provide seamless multi-layer-data set maps and information relevant to fire management and fuels treatment decisions.) Details of the LANDFIRE data tool are available at: http://www.landfire.gov/participate_refdata_sub.php. The tool includes spatial data from several federal agencies, state governments, municipalities, academic institutions, and others.

[19] There is no readily available list of databases pertaining to water quality, though individual agencies maintain information about some databases. See, for example, USDA's information on data that pertains to water and agriculture at: http://wqic.nal.usda.gov/databases-0, which lists data sources for USDA, USGS, and EPA. Other agencies also maintain water quality databases, including, for example, the Army Corps of Engineers, FWS, Tennessee Valley Authority, and NOAA. Reporting on data integration and other water management challenges, see U.S. Army Corps of Engineers. 2010. Responding to National Water Resources Challenges. Washington, DC: U.S. Army Corps of Engineers, Civil Works Directorate.

[20] Cloud computing is defined by the National Institute of Standards and Technology (NIST) as "a model for enabling convenient, on-demand network access to a shared pool of configurable computing resources (e.g., networks, servers, storage, applications, and services) that can be rapidly provisioned and released with minimal management effort or service provider interaction" (Kundra, 2011). See Mell, P., and T. Grance. 2011. The NIST Definition of Cloud Computing. NIST Special Publication 800-145. Online. Avail-

the potential to play a major role in addressing these issues and improving government operational efficiency (Kundra, 2011) by helping agencies that are grappling with the need to provide highly reliable, innovative services quickly despite resource constraints.

A similar fragmentation frequently happens with basic and applied research. While there is some coordination among agencies in constructing research portfolios and ensuring that results are available to all participating agencies[21], individual agencies generally undertake research within their silos, directed at meeting their needs or tailored to their programs. Again, this approach enhances expertise and accountability, but it frustrates the initiation of cross-agency research, even for shared domains. Also, as with funding requirements, information gathering for sustainability purposes should extend beyond the short term. Indeed, research needed to understand the connections among domains and the mechanisms that sustain critical functions often requires studies that extend for decades, rather than the typical three-year grants given by research agencies. A notable exception to short-term research relevant to sustainability is the National Science Foundation's (NSF's) Long-Term Ecological Research (LTER) program; however, it represents only a modest investment relative to the scale of the need. Further long-term research investments grounded in practical questions and knowledge gaps identified by decision makers are needed.

CULTURE OF GOVERNMENT

A discussion of the impediments or barriers to successful governance linkages must also include the nature of government service or what is often referred to as the "culture of government." This is a very broad and well-researched subject on which a great deal has been written (Rainey, 2009; Kettl, 2009a; Kettl, 2009b; McKinney et al., 2010; Bardach, 1998; Daley, 2009). The starting point is typically an acknowledgement that the underlying principles of good government—what is expected of government employees—are themselves in tension: specialization versus integration; certainty versus adaptive management; and uniformity versus flexibility (see Box 2-1).

For our purposes, one of the more prevalent characteristics of government service flows in part from the silo phenomenon—namely, agencies and their personnel tend to go it alone. To be sure, many agencies recognize the importance of collaboration and are trying to coordinate in appropriate circumstances. A number of tools for consultation and collaboration have been devel-

---

able at http://csrc.nist.gov/publications/drafts/800-145/Draft-SP-800-145_cloud-definition. pdf. Accessed February 25, 2013.

[21]An example of a success story in this realm is the Tox21 collaboration http://epa.gov/ncct/Tox21, which features well coordinated research, shared across programs, with results made publicly available.

oped and are being deployed.[22] Similarly, there are a number of leaders (at both the senior and staff levels, as we will see in Chapter 5) who are risk takers or innovators. Nonetheless, the prevailing view is that such efforts are often very difficult, much more difficult than they should be. Simply stated, for the most part, agency personnel tend to focus on the agency's statutory mandates and proceed in the conventional way. Recognition, promotion, and other rewards are all based on advancing the agency's agenda in a competent but orthodox fashion.[23] Risk aversion is the norm, for if something goes wrong when taking initiative or doing something unusual, there are likely to be adverse consequences from senior administration officials, congressional oversight, regulated entities, and/or the press; the notion that failure can be beneficial in that people can learn from mistakes is not something that has salience in the world of government. Not surprisingly, therefore, training is traditionally focused on existing agency practices and processes rather than adaptive management, collaboration, or other efforts to innovate and integrate actions across agencies.

BOX 2-1
Governance and Institutional Design

All governance models—of whatever scale or purpose—are ultimately judged by four criteria: legitimacy, fairness, effectiveness, and efficiency. Yet each of these features may be achieved through a number of procedural and structural characteristics, which are themselves often in tension with one another. These tensions may be particularly evident in efforts to govern through networks in contexts that involve multi-issue integration, complex connections, multiple decision makers, and uncertainties about present or future conditions and how governing will affect those conditions. Legitimacy requires uniform application of laws, for example, while effectiveness and efficiency may require flexibility and adaptability, both of which may be in tension with uniformity. Effectiveness requires clear accountability, but fairness in a context of issues affecting intergovernmental partners and the private sector may require diffused and shared responsibilities. There is the perennial tension between timely decision making and stakeholder inclusivity and involvement. With respect to structural and procedural characteristics, there may be tension between specialization and multi-issue integration, or between expertise and accessibility. Similarly, there may well be tension between the goal of predictability or certainty on the one hand, and the importance of adaptability or resilience on the other.

---

[22]Executive Office of the President Office of Management and Budget and Council on Environmental Quality. 2012. Memorandum on Environmental Collaboration and Conflict Resolution. Online. Available: http://energy.gov/sites/prod/files/OMB_CEQ_Env_Collab_Conflict_Resolution_20120907-2012.pdf. Accessed December 10, 2012.

[23]One manifestation of this is in the implementation of some governmentwide management statutes such as the Government Performance Results Act, in which managers have tended to identify goals that are agency-specific, or even program-specific, rather than more integrated ones; and it is well established that people do what is being measured.

Another reason collaboration is rarely emphasized is because part of the government culture is to "stay in your lane"—that is, work your area of responsibility and avoid getting involved in a sister agency's activities. If there is no budget, there is no cover or incentive for going beyond the established boundaries and eliciting the charge of "mission creep." In other words, lanes and silos have much in common. There are exceptions to this rule, especially in the interagency processes conducted by OMB (such as in the realm of regulations), but these are clearly the exception and generally require some senior official in the Executive Office of the President to act as a convener or honest broker (or both) if interagency approaches are to be effective.

In sum, a number of impediments or barriers frustrate federal government efforts to create linkages to address sustainability issues. The structural fragmentation, funding constraints, lack of coordination of information and research, and the culture of government are simply not conducive to partnerships or to extensive collaboration with other affected or invested entities. Because many sustainability issues cross agency boundaries and require long-term investment, these situations create challenges to effective government response. But these challenges can be overcome, as we will discuss in the following chapters.

REFERENCES

Acquisition Central. 2007. Federal Acquisition Circular. No. 2005-17. Online. Available at https://www.acquisition.gov/far/fac/FAC%2017%20Looseleaf_r.pdf. Accessed November 5, 2012.

Bardach, E. 1998. Getting Agencies to Work Together. Washington, DC: Brookings Institution.

Bingham, L. B. 2010. The Next Generation of Administrative Law: Building the Legal Infrastructure for Collaborative Governance. Wisconsin Law Review 1:297-356.

CEF (Corporate Eco Forum). 2012. About CEF. Online. Available at http://www.corporateecoforum.com/contact/index.php. Accessed October 29, 2012.

Daley, D. 2009. Interdisciplinary problems and agency boundaries: Exploring effective cross-agency collaboration. *Journal of Public Administration Research and Theory* 19 (3):477-493.

DOI (U.S. Department of the Interior), Office of Inspector General. 2007. Proper Use of Cooperative Agreements Could Improve Interior's Initiatives for Collaborative Partnerships. Report No. W-IN-MOA-0086-2004. Online. Available at http://www.gpo.gov/fdsys/pkg/GPO-DOI-IGREPORTS-2007-g-2005/pdf/GPO-DOI-IGREPORTS-2007-g-2005.pdf. Accessed November 5, 2012.

DOI. 2012. Landscape Conservation Cooperatives. Online. Available at http://www.doi.gov/lcc/index.cfm. Accessed November 5, 2012.

EPA (U.S. Environmental Protection Agency). Summary of the Clean Air Act. 42 U.S.C. §7401 et seq. (1970). Online. Available at http://www.epa.gov/lawsregs/laws/caa.html.

EPA. 2013. Tox 21. Online. Available at http://epa.gov/ncct/Tox21. Accessed April 19, 2013.

Executive Office of the President Office of Management and Budget and Council on Environmental Quality. 2012. Memorandum on Environmental Collaboration and Conflict Resolution. Online. Available: http://energy.gov/sites/prod/files/OMB_

CEQ_Env_Collab_Conflict_Resolution_20120907-2012.pdf. Accessed December 10, 2012.

HHS (Department of Health and Human Services). 2013. National Toxicology Program. Online. Available at http://ntp.niehs.nih.gov/?objectid=7201637B-BDB7-CEBA-F57E39896A08F1BB. Accessed April 19, 2013.

Kagan, E. 2011. Presidential Administration. 114 *Harvard Law Review* 2245:2261-2262.

Kaufman, D. 2012. Forging Interagency Linkages on Sustainability: Federal Emergency Management Agency's (FEMA) Perspective. Presentation to the National Research Council Committee on Sustainability Linkages in the Federal Government 6th Meeting. October 11, 2012.

Kettl, D. F. 2002. The Transformation of Governance: Public Administration for Twenty-first Century America. Baltimore, MD: Johns Hopkins University Press.

Kettl, D. F. 2009a. Administrative Accountability and the Rule of Law. *Political Science and Politics* 42(1):11-17.

Kettl, D. F. 2009b. The next government of the United States: Why our institutions fail us and how to fix them. New York, NY: Norton & Co.

Kundra, V. 2011. Federal Cloud Computing Strategy. Online. Available at http://www.dhs.gov/sites/default/files/publications/digital-strategy/federal-cloud-computing-strategy.pdf. Accessed February 25, 2013.

LANDFIRE. 2013. Contribute Data. Online. Available at http://www.landfire.gov/participate_refdata_sub.php. Accessed April 19, 2013.

McKinney, M, L. Scarlett, and D. Kemmis. 2010. Large Landscape Conservation: A Strategic Framework for Policy and Action. Cambridge, MA: Lincoln Institute of Land Policy.

Mell, P., and T. Grance. 2011. The NIST Definition of Cloud Computing. NIST Special Publication 800-145. Online. Available at http://csrc.nist.gov/publications/drafts/800-145/Draft-SP-800-145_cloud-definition.pdf. Accessed February 25, 2013.

National Archives. Federal Advisory Committee Act (5 U.S.C. Appendix 2). Online. Available at http://www.archives.gov/federal-register/laws/fed-advisory-committee. Accessed November 5, 2012.

National Science Foundation. 2011. Long-Term Ecological Research Program: A Report of the 30 Year Review Committee. Online. Available at http://portal.nationalacademies.org/portal/server.pt/gateway/PTARGS_0_412321_4829_954_425760_43/collab/docman/download/337538/0/0/0/LTER-A%20Report%20of%20the%2030%20Year%20Review%20Committee.pdf. Accessed October 4, 2012.

Nidumolu, R., C. K. Prahalan, and M. R. Rangaswami. 2009. Why sustainability is now the key driver of innovation. *Harvard Business Review* 87.

OMB (Office of Management and Budget). 1978. Implementation of the Federal Grant and Cooperative Agreement Act of 1977. *Federal Register* 43(161):36860-36865. Online. Available at http://www.commerce.gov/sites/default/files/documents/2011/september/omb-grants-and-contracts-guide-1978.pdf. Accessed November 5, 2012.

OMB. Circulars. Available at http://www.whitehouse.gov/omb/circulars_default. Accessed November 5, 2012.

OMB. Government Performance and Results Act (GPRA) Related Materials. Online. Available at http://www.whitehouse.gov/omb/mgmt-gpra/index-gpra. Accessed November 5, 2012.

Orszag, P. R. 2009. Memorandum for the Heads of Executive Departments and Agencies. Subject: Open Government Directive. Online. Available at http://www.whitehouse.gov/open/documents/open-government-directive. Accessed November 5, 2012.

Pressman, J. L., and A. Wildavsky. 1973. Implementation. Berkeley, CA: University of California Press.

Rainey, H. G. 2009. Understanding and Managing Public Organizations. San Francisco, CA: John Wiley & Sons.

Resource Conservation and Recovery Act of 1976. 42 U.S.C. 6901 et seq. Online. Available at http://el.erdc.usace.army.mil/emrrp/emris/emrishelp5/resource_conserva tion_and_recovery_act_legal_matters.htm. Accessed November 1, 2012.

Sustainable Agriculture Research and Education. 2013. Online. Available at http://www. sare.org. Accessed March 12, 2013.

The White House. 2000. Executive Order 13148—Greening the Government through Leadership in Environmental Management. *Federal Register* 65(81):24607-24611.

The White House. 2009. Executive Order 13514—Federal Leadership in Environmental, Energy, and Economic Performance. *Federal Register* 74(194):52117-52127.

U.S. Army Corps of Engineers. 2010. National Report: Responding to National Water Resources Challenges. Washington, DC: U.S. Army Corps of Engineers, Civil Works Directorate.

U.S. Court of Appeals for the District of Columbia Circuit. 2012. No. 11-1302, August 21, 2012. EME Homer City Generation, L.P., Petitioner v. Environmental Protection Agency, et al., Respondents. Online. Available at http://www.cadc.uscourts. gov/internet/opinions.nsf/19346B280C78405C85257A61004DC0E5/$file/11-1302-1390314.pdf. Accessed November 5, 2012.

USDA National Agricultural Library. Water Quality Information Center Databases. Online. Available at http://wqic.nal.usda.gov/databases-0. Accessed November 5, 2012.

U.S. House of Representatives. Federal Grant and Cooperative Agreement Act of 1977 (31 U.S.C. 6301 et seq.). Online. Available at http://uscode.house.gov/download/pl s/31C63.txt. Accessed November 5, 2012.

U.S. Senate. 2002. Endangered Species Act of 1973 [As Amended Through Public Law 107-136, Jan. 24, 2002]. Online. Available at http://epw.senate.gov/esa73.pdf. Accessed November 5, 2012.

U.S. Senate. 2002. Federal Water Pollution Control Act (33 U.S.C. 1251 et seq.). Online. Available at http://epw.senate.gov/water.pdf. Accessed November 5, 2012.

U.S. Senate. 2004. The Clean Air Act [As Amended through P.L. 108-201, February 24, 2004]. Online. Available at http://epw.senate.gov/envlaws/cleanair.pdf. Accessed November 5, 2012.

U.S. Senate Committee on Appropriations Subcommittee on Energy and Water Development. Online. Available at http://www.appropriations.senate.gov/sc-energy.cfm. Accessed November 5, 2012.

U.S. Senate Committee on Appropriations Subcommittee on Interior, Environment and Related Agencies. Online. Available at http://www.appropriations.senate.gov/sc-interior.cfm. Accessed November 5, 2012.

U.S. Supreme Court. 1988. Bowen v. Georgetown Univ. Hosp., 488 U.S. 204 (1988) at 208. Online. Available at http://caselaw.lp.findlaw.com/scripts/getcase.pl?court= us&vol=476&invol=355. Accessed October 1, 2012.

Chapter 3

Examples of Sustainability Connections and Linkages

The committee held a series of fact-finding meetings to explore six cases that posed challenges in terms of connected resources—challenges in the areas of science, monitoring, organization, and governance—and to examine the approaches various agencies used to address them. Although they differed substantially, each example dealt with the economy, society, and environment: the "three-legged stool" of sustainability. In many cases, stakeholder agencies and organizations, working with relevant government agencies, achieved significant and sustainable results because the right people with the right approaches from disparate organizations came together to do so. An agenda for each workshop is provided in Appendix C.

URBAN SYSTEMS - Philadelphia

For much of the latter half of the 20th century, Philadelphia, a city of 1.5 million, was a city in decline. In 2007 mayoral candidate Michael Nutter adopted sustainability as the central organizing principle of his campaign, envisioning a revitalized Philadelphia as the number one "Green City" in America.

This vision resonated with the public, and once elected, Mayor Nutter issued a citywide sustainability plan: "Greenworks Philadelphia." The plan considered sustainability through five lenses: energy, environment, equity, economy, and engagement. Five goals and 15 measurable targets were designed to be achieved by 2015. Public perception depended upon measuring progress and communicating it in a compelling way. The city continues to measure and publicize results, both positive and negative.

Philadelphia is one of four U.S. metropolitan areas where the U.S. Department of Transportation (DOT) has established an office to plan and manage both public transit programs and highways in the metropolitan region. In addition, just last year the Natural Resources Defense Council recognized Philadelphian efforts to implement innovative transportation policies and practices by taking a complete streets approach, increasing the availability of walking and bicycling trails and improving public transit.

Similarly, several innovative initiatives have converted vacant lots in the city to parks and other green spaces, which have been shown to improve the health and safety of those nearby (Branas, 2012). For example, a program was developed to remove trash and debris from vacant lots, grade the land, plant grass and trees to create parklike settings, and install low wooden post-and-rail fences. During the fourth committee meeting held in June 2012, Charles Branas, University of Pennsylvania, reported that these greening efforts had positive, significant impacts on several health outcomes: Gun assaults were reduced in all city sections; vandalism dropped in West Philadelphia; high stress decreased among residents in North Philadelphia; and exercise increased among residents of West Philadelphia.

The Clean Water Act of 1972 prescribes that local governments capture and treat wastewater before discharging it in rivers. Most cities treat sewage and runoff separately; however, portions of many older cities, such as Philadelphia, collect both sewage and runoff in the same system. When the combined volume exceeds the capacity of the sanitary system, the excess is discarded into the nearby river—discharges that must be minimized under the Clean Water Act. This problem, Combined Sewer Overflow (CSO), has been exacerbated by urban build-out; as more green space is paved, there is more runoff. The U.S. Environmental Protection Agency (EPA) requires cities to address CSO, in most cases by expanding their sanitary sewer or via separate piping and treatment systems. The CSO infrastructure is expensive, and there are ongoing costs.

Philadelphia took the novel approach of reusing or managing rainwater in order to prevent runoff. The alternative to CSO, called "Green Stormwater Infrastructure" (GSI), allows stormwater to percolate through the soil wherever possible, using devices such as tree trenches, wetlands, planters, green roofs, pervious paving, or rain gardens (Figure 3-1). Rain barrels promote recovery and recycling of water.

The linkages between the GSI and other systems—and the benefits that result—are not obvious, but they are remarkable (PWD, 2011):

- **Saving energy while mitigating and offsetting climate change**. Trees and plants are an important part of the GSI, shading and insulating buildings from wide temperature swings and decreasing the energy needed for heating and cooling. Because rain is managed where it falls in systems of soil and plants, energy is not needed for traditional systems to store, pump, and treat it. Growing trees also act as carbon "sinks," absorbing carbon dioxide from the air and incorporating it into their branches and trunks.
- **Restoring ecosystems**. Allowing rain to soak into the ground and return slowly to streams restores a water cycle similar to that of a natural watershed. Soil is a natural water filter, and this percolation limits erosion of stream

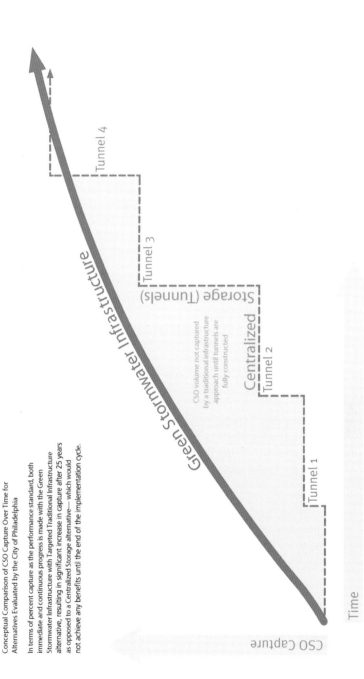

FIGURE 3-1 An Unconventional Path: Rationale for the Green Infrastructure Approach. SOURCE: Presentation by Christopher Crockett, Philadelphia Water Department (PWD), June 12, 2012.

channels caused by high flows. This approach includes physical restoration of stream channels and streamside lands, including wetlands, to restore habitat for aquatic species. Soil also filters storm water runoff, reducing pollutant concentrations and improving surface and groundwater quality.

• **Conserving water.** Rainwater reuse technologies collect, convey, and store rain from relatively clean surfaces such as roofs. The water is generally stored in a tank or cistern and then reused for irrigation or flushing toilets in residential properties, and for boilers or cooling towers in industrial or commercial properties. If the cisterns store water for predictable, year-round use, their use can count toward compliance with stormwater regulations and also save water.

• **Reducing health effects of excessive heat**. Heat waves are a fixture of summers in Philadelphia, including some severe enough to result in premature deaths. Trees and green roofs that are part of GSI reduce the severity of extreme heat events in three ways: by creating shade, by reducing the amount of heat-absorbing pavement and rooftops, and by emitting water vapor—all of which cool hot air.

• **Enhancing recreation**. Throughout the park system, impervious cover such as concrete pavement is reduced, and residents enjoy recreation along Philadelphia's stream corridors and waterfronts.

• **Improving air quality, with benefits for health**. Like many major cities in the United States, EPA currently classifies the Philadelphia metropolitan area as exceeding federal air quality standards. GSI's expansion of green areas improves air quality because it can lead to lower air temperatures, which in turn reduce smog formation. Green areas can absorb air pollutants and lower carbon dioxide levels.

• **Increasing property values**. Trees and parks contribute to making an urban neighborhood an inviting place to live, work, and play. Residents clearly recognize and value the quality-of-life benefit of urban vegetation. Property values are higher close to parks and greenery.

• **Improving safety**. Alongside the GSI program, the Pennsylvania Horticultural Society (PHS) and other community and municipal partners promoted a program to clean, green, and maintain abandoned vacant lots—a process that produces visually pleasing results and, surprisingly, adds to public safety. Vacant lots are known to be convenient places to store weapons. Greening them reduced such storage and correspondingly reduced gun assaults in the neighborhood.

The overall effort was possible not only because the commitment to sustainability was adopted and supported at the highest level, but also because of the leadership and expertise of key actors from the Philadelphia Water Department (PWD). In 2006, PWD changed its regulations to require that all new construction projects in the city infiltrate, detain, or treat on-site the first inch of rainwater. This approach included charging nonresidential land owners for stormwater control based upon parcel characteristics. The new fee formula called for 80 percent of a property's charges to be based on the amount of its impervious surface area and 20 percent on its gross area. This principle was later extended to building regulations to encourage "green roofs."

Through local experience and networks, linkages were created with other city departments, and benefits other than simple stormwater management were achieved. The plan came together when it was agreed that water revenue dollars could be spent on solutions that achieved not only the intended water pollution control but also other benefits.

As discussed above, in the case of sustainability efforts in Philadelphia, success was driven by the clear vision and commitment shown by leaders and supported by the innovation and dedication of technical experts and champions in the field to implement performance-driven standards (see Box 3-1 for sustainability performance outcomes related to the sustainability initiatives described above). Well-developed communication elements were also critical to the success.

URBAN SYSTEMS - Phoenix

Phoenix is a large, rapidly growing city located in a desert environment with an ethnically diverse and rapidly expanding population. Phoenix faces a unique combination of sustainability challenges, including water scarcity, poor urban air quality, significant loss of biodiversity, increasing demands on energy resources, and urban heat island effects on public health. The changing climate may exacerbate some of these challenges and increase the importance of addressing them in a timely manner to sustain quality of life for Phoenix residents.

In 2009 former Phoenix Mayor Phil Gordon put forward his Green Phoenix plan. Gordon's vision included increasing the use of solar energy and improving transportation projects to make Phoenix the first carbon-neutral city in the country. The city has strong linkages to national groups (U.S. Mayors, EPA training programs, U.S. Forest Service, nongovernmental organizations (NGOs), State of Arizona Department of Environmental Quality), local communities, and corporations. However, despite the importance of these efforts, the mayor's office currently has only one person assigned the role of sustainability advisor, and that person has no designated budget authority.

Water and land use

With an average annual rainfall of only 20 cm, water availability and quality are pressing issues for Phoenix residents. Rainfall is highly variable from year to year, making water use planning difficult. The Arizona water supply is currently divided between surface water from the Colorado, Salt/Verde, and Gila River systems (54 percent); five major groundwater aquifers (43 percent); and

---

BOX 3-1
Sustainability Performance Outcomes: Philadelphia Example

Greening Public Lots (Source: Branas et al., 2011; Branas, 2012)
Purpose: Removes trash and debris; grade the land; plant grass and trees to create a park-like setting; install low wooden post-and-rail fences.
Goal: Spur economic development.
Major Players: Funding received from PHS, City of Philadelphia, CDC, and NIH
Metrics: Gun assaults, vandalism, stress, exercise, substance use.

Septainable (Source: SEPTA, 2011)
Goals: Develop a more competitive transit system and attractive mobility alternative; improve environmental stewardship and build livable communities; increase economic prosperity across Greater Philadelphia
Major Players: Southeastern Pennsylvania Transportation Authority (SEPTA), City of Philadelphia, DOT, the Department of Housing and Urban Development (HUD), EPA
Metrics: Greenhouse gas and air pollutants; water usage; waste; farmers' markets; transit-oriented development; infrastructural improvements; transit mode increases.

Greenworks Philadephia (Source: City of Philadelphia, 2009)
Goal: To create Philadelphia the greenest city in America
Major Players: City of Philadelphia; numerous agencies, universities, foundations
Metrics: Energy consumption; greenhouse gas emissions; waste; parks; limit food deserts; reduce vehicle miles traveled by 10 percent.

Green Stormwater Program (Source: PWD, 2011)
Goals: Reduce runoff; update water and sewer system; come into compliance with federal and state laws.
Benefits: Improve air quality; save energy; restore ecosystems; reduce social cost of poverty; enhance recreation and quality of life; reduce effects of excessive heat.
Major Players: PWD, City of Philadelphia, EPA, State of Pennsylvania
Metrics: Implement intensive large-scale application of GSI; increase wet weather wastewater treatment capacity in targeted locations.

---

3 percent effluent. Agriculture is the largest user of water in Arizona; however, this may decline in the future given predictions that the climate of the southwestern United States will become drier over this century (Overpeck and Udall, 2010). To optimize water use, Phoenix will need long-term planning horizons that incorporate uncertainty and trade-offs (Quay, 2012; Arizona State University Morrison Institute for Public Policy, 2011).

The Maricopa Park plan is a Phoenix-based example of local and federal linkages, fostering conservation of species, preservation of habitat, and recreational opportunities. Maricopa County Park systems ring the city of Phoenix with 163,000 acres of desert mountain preserves, constituting the largest set of wild land preserves in any metropolitan region in the United States, although the urban environment is rapidly enveloping those areas (McCue, 2012). The Conservation Alliance was formed to preserve sustainable lands within Phoenix; partners include Arizona State University (ASU), city and county governments, private foundations, and NGOs. Federal agencies contribute to this effort through "America's Great Outdoors," which involves the U.S. Department of the Interior (DOI), the Department of Agriculture (USDA), the Department of Defense (DOD), the Department of Commerce, and the Council on Environmental Quality (CEQ).

Importance of foundational science

Long-term research and its application are critical to understanding, promoting, and enhancing sustainability in urban environments. Phoenix is the site of one of only two urban programs supported by the National Science Foundation's (NSF's) Long-Term Ecological Research (LTER) program, the Central Arizona-Phoenix (CAP) LTER, which aims to integrate biological, ecological, engineering, economic, and social sciences. CAP LTER research has several integrative focus areas including climate, ecosystems, and people; water dynamics in a desert city; biogeochemical patterns, processes, and human outcomes; and human decisions and biodiversity. The CAP LTER is led by ASU, which works with a wide range of community partners through its School on Sustainability to achieve its research, education, and outreach goals.

Urban heat islands and sustainability of healthy populations

In a desert city such as Phoenix, the urban heat island effect can be very pronounced in two ways. First, in the summer months, some urban areas may be several degrees hotter than others. Within the city, microclimates exist in neighborhoods as a function of vegetation and its effect on cooling by evaporative transpiration; in fact, local temperatures within Phoenix can vary by as much as 14 degrees F (Harlan et al., 2006). Second, the urban heat island effect is reflected in generally higher *minimum* daily temperatures because heat is retained by the built environment, which fails to cool at night as the surrounding desert does (Figure 3-2).

One focus area is on the public health impacts of the urban heat island effect. This foundational research brings together climatologists, ecologists, sociologists, geographers, and geoscientists. Research reveals that vulnerability to extreme heat depends on place and social context. The highest morbidity and mortality associated with extreme heat falls disproportionately upon marginalized groups, including the poor, minorities, and the elderly (Harlan et al., 2006). Substandard housing, lack of air conditioning, crowding, poverty, homelessness, and aging contribute to the occurrence of heat-related health problems, as do

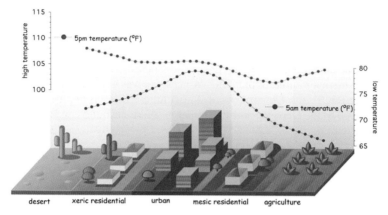

FIGURE 3-2 Urban Heat Island expression in Phoenix. SOURCE: Presentation by Diana Petitti, ASU, June 11, 2012. The diagram was created by Joseph Zehnder and Susanne Grossman-Clarke, Central Arizona – Phoenix Long-Term Ecological Research project, ASU.

certain occupations such as construction and agriculture. A warming climate will undoubtedly exacerbate heat-related health problems, particularly in urban environments. Harlan (2012) cites several benefits of improving ecosystem services with increased vegetation in heat-vulnerable neighborhoods, such as psychological impacts, including reducing stress; promoting health and well-being; providing recreational spaces for outdoor physical activities; improving air quality; and other health-related outcomes including reducing the number of heat-related illnesses.

Federal linkages in support of sustainability

During the fourth committee meeting held in June 2012, Petitti (2012) described several positive examples of linkages among agencies and organizations working to address sustainability issues in Phoenix:

• The Centers for Disease Control and Prevention (CDC) is supporting an Arizona Department of Health Services grant related to preparedness for high heat events.

• USDA is providing support to local groups for urban agriculture programs, such as community gardens in poor neighborhoods, which provide green spaces that cool the environment and reduce social isolation.

• The National Weather Service and the State of Arizona are collaborating to provide heat watch warnings and public education.

• State and local agency employees and nonprofits have formed the Coalition on Heat Relief, which focuses on protecting the homeless by passing out water and getting people out of the heat in the summer.

Thus, a limited number of federal agencies are effectively engaging with state and local governments and other organizations to address the problem of public health impacts of urban heat islands.

This effort to improve human health and well-being integrates many aspects of sustainability, including greater availability of energy, water, green space, and transportation; improvements in air quality; support for social equity; and adaptation to a changing climate. Two overall conclusions may be drawn from this example. First, additional federal partners, including some unobvious ones, need to be engaged in supporting urban sustainability. Housing, transportation, and energy are critically important to populations vulnerable to extreme heat events. Partnerships and a shared vision for urban sustainability among federal, state, and local governments and organizations, with clearly articulated roles and responsibilities, can reduce the need for crisis management and last minute interventions.

Second, the contributions made by CAP LTER show that long-term, integrative, cross-disciplinary research provides a strong scientific foundation for decision making. Over 80 percent of the U.S. population lives in or near a city, and yet there is little long-term research on urban sustainability. Additional federal science funding agencies must step forward to support this important endeavor.

NONURBAN SYSTEMS - Mojave Desert

The Mojave Desert in California is a vast and seemingly harsh, yet fragile region; however, despite common perception, the desert is far from empty. The land is used for recreation, housing, and military training. It is a premium location for renewable energy development, as it has some of the highest-quality solar and wind resources in the nation. It is also home to mining, agriculture, energy production, and a wide variety of human and natural communities, as well as unique ecosystems and a number of endangered species. The competition between human-centric land uses and the desire to preserve species habitat and manage on an ecosystemwide basis has increased the need for coordinated land management in the Mojave Desert.

The desert is largely public land overseen by a patchwork of organizations. In the California Mojave, approximately 80 percent (25 million acres) of the land is publicly owned, including two national parks, one national preserve, 72 wilderness areas, six military bases, and 10 state parks. In addition, the area involves eight county jurisdictions and 37 federally recognized Native American Indian tribes. Conflicting demands for the use of California desert lands make it imperative that governmental agencies cooperate to support agency missions, protect desert resources, and manage public use. Land management in the California Mojave currently involves two coordinated management efforts: the Desert Manager's Group (DMG) and the ongoing development of the Desert Renewable Energy Conservation Plan (DRECP).

Desert Managers Group

This group originated as a collaboration between the Bureau of Land Management (BLM) and the National Park Service (NPS) in response to the impending passage of the California Desert Protection Act. That Act and the vicissitudes of history have resulted in major transfers of land from BLM to NPS, large wilderness designations, and responsibilities to administer adjacent and sometimes overlapping pieces of land (Figure 3-3).

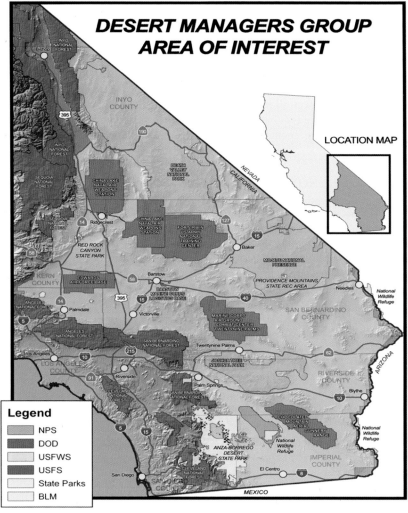

FIGURE 3-3 DMG Area of Interest. SOURCE: Presentation by Russell Scofield, U.S. Department of the Interior, April 11, 2012.

As the largest manager of public land in the Mojave Desert, BLM was engaged in ecosystem-based planning efforts, and the land transfers to NPS created conflict between the organizations over land-use planning and management. There were major conflicts over grazing, desert tortoise recovery, off-road vehicle use, mining, hunting, military overflights, burro and wild horse management, water for wildlife, and development (NPS, 2003).

In late 1994, the Desert Managers Group was officially established to include NPS, BLM, U.S. Fish and Wildlife Service (FWS), California State Parks, and the four military base commands. The group allows agencies to avoid duplication of effort by sharing information and facilities; it also coordinates federal efforts to acquire lands, eliminating situations where multiple agencies bid on the same parcel, which would artificially inflate property values.

DMG is enhanced by longstanding personal and professional relationships among the participants. Regular face-to-face meetings allow line officers to work as a group, set goals, address cross-boundary issues and get to know each other better. Informal networking is an important component.

Although the function of DMG requires support from senior leadership, its success depends upon creating value for the participants. One dimension of this value is derived from the enhanced decision space they gain by participating in DMG. The group causes the individual members to think outside their own organizational boundaries and to enlarge the interpretation of their own agency mission to focus on landscape issues and regional sustainability. Bimonthly meetings hosted by the group focus on land management issues of common concern. Although DMG is a very successful collaboration, it has neither budget nor regulatory authority over land use or other sustainability-based decisions, and thus its impact is limited to coordination of voluntary efforts by its members.

Desert Renewable Energy Conservation Plan

In 2008 California Governor Arnold Schwarzenegger signed Executive Order S-14-08, which requires that one-third of California's energy come from renewable sources by 2020. In response, the California Energy Commission (CEC), California Department of Fish and Game (CDFG), FWS, and BLM signed a Memorandum of Understanding to expedite the permitting process for renewable energy projects, including those on federally owned land in the Mojave Desert region.

The executive order also requires the development of a Natural Communities Conservation Plan (NCCP): a cooperative effort to protect habitats and species authorized under the NCCP Act of 2003. The primary objective of the NCCP program is to conserve natural communities at the ecosystem level while accommodating compatible land use. In the California context, this NCCP is known as the Desert Renewable Energy Conservation Plan (DRECP). The DRECP will also produce a habitat conservation plan to comply with the Federal

Endangered Species Act (ESA) and a land use plan amendment in accordance with the Federal Land Policy and Management Act.

The DRECP process is led by the Renewable Energy Action Team (REAT), which is a collaboration of state and federal agencies, including those mentioned above. The DRECP includes a Stakeholder Committee and Science Advisory process, and its planning horizon is 25 to 40 years. The DRECP will be implemented through specified conservation, avoidance and minimization measures, and a science-based monitoring and adaptive management program.

Coordinated Management Assessment

The DRECP and DMG involve many common stakeholders, but their purpose and total membership is different. The DRECP is an example of a state–federal collaborative planning process, involving a much wider group of participants and stakeholders than DMG. DMG was chartered to facilitate ongoing collaboration among members in a closed forum. The DRECP has a clearly defined, finite process and regulatory authority over land use for one type of activity—renewable energy projects. While the DRECP has likely benefitted from the existence of DMG, the two processes are not formally connected.

Both the DRECP and DMG are efforts to collaborate across levels of government and agency responsibilities. They were initiated because holistic solutions to complex problems involving energy development, ecosystem conservation, and the public interest could not be developed and implemented by one agency or one governance level. These efforts to achieve sustainable solutions are works in progress, but they are vivid examples of the links needed to achieve those solutions. As this example illustrates, reaching sustainability goals requires partnerships that move beyond traditional organizational boundaries. These partnerships can allow for the coordination of activities and sharing of critical information. Also, employing adaptive approaches can allow for flexibility in anticipating new challenges to address complicated sustainability linkages.

NONURBAN SYSTEMS - Platte River

The Platte River flows through three states, irrigates 2.8 million acres, generates 400 MW of electric power, provides water to 2.5 million people, and supports significant wildlife habitat (Freeman, 2010; Figure 3-4). After a decade of negotiations about how to protect endangered species along the Platte while maintaining the river's usefulness for irrigation and other purposes, the Platte River Recovery Implementation Program was initiated in 2007 through the "Cooperative Agreement for Platte River Research and Other Efforts Relating to Endangered Species Habitat along the Central Platte River, Nebraska" (Platte River Recovery Implementation Program, 2010). The Cooperative Agreement and related documents, signed by the governors of Nebraska, Wyoming, and

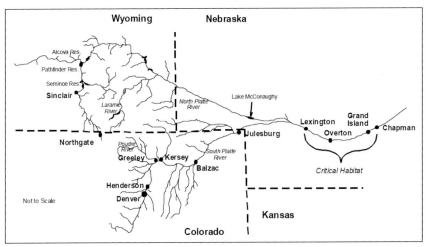

FIGURE 3-4 The Platte River Basin. SOURCE: Presentation by David M. Freeman, Colorado State University, April 12, 2012. Reprinted with permission from the University Press of Colorado.

Colorado and by the Secretary of the Interior, establish a shared vision and responsibility for managing the central Platte River. Through the agreement, participants created: 1) a governance structure with which to coordinate decisions and actions among three states, several federal agencies, special districts, power providers, water managers, and the agricultural sector; and 2) a decision framework that links species protection, groundwater and surface water management, power production, and land management.

Though the Platte River Program takes into account multiple water and land uses, it was created in response to statutory responsibilities to protect endangered species (Freeman, 2010). The program sets forth provisions for implementing certain aspects of FWS's recovery plan for four endangered or threatened species along the Platte River in a context where water is used for multiple purposes in a predominantly agricultural region.[1] Specific elements of the program include: (1) recovering more historical patterns of stream flow during relevant times of the year through re-timing of flows and water conservation and supply projects, and (2) enhancing, restoring, and protecting habitats for the four listed species (Scarlett, 2012). The program recognizes the interconnections between groundwater and surface water management. The program's implementation prompted passage of a groundwater management statute in Nebraska to provide the state with one tool necessary to achieve its program goals.

---

[1] The four species include the endangered whooping crane, least tern, and pallid sturgeon and the threatened piping plover.

Developing and implementing actions to protect the four endangered species required changes in both land and water management and presented questions about the magnitude, timing, frequency, and temperature of water flows. Scientific uncertainties and complexities made it difficult to determine appropriate actions to address the species' needs. A key component of the overall Platte River Program—one that helps address these uncertainties—is an Adaptive Management Plan, which provides an iterative process to test hypotheses about management strategies that will most closely achieve program objectives. Monitoring for improvements in river form and function, as well as in the status of the four species, guides decisions about the most appropriate management strategies.

The Cooperative Agreement established a Governance Committee (GC) as the decision-making body for the Platte River Program (Platte River Recovery Implementation Program, 2007). The GC has ten members, representing the three watershed states, two federal agencies (FWS and BOR), water users from each of the three states, and representatives from two environmental organizations. Though all key stakeholders participate, funding is provided by the three states and BOR. Initially, the GC guided a planning process that culminated in a Final Program Agreement signed by the three governors and the Secretary of the Interior in January 2007 (Scarlett, 2012). At the third committee meeting, held in April 2012, presenter David Freeman of Colorado State University noted that the multiparticipant GC operates with clear decision rules that require 9 of 10 members to concur on any major policy decision and 7 of 10 to agree on nonpolicy issues (Freeman, 2012).

The GC, which is responsible for implementing the program, contracted with a private natural resources consulting firm, Headwaters Corporation, to provide ongoing program management. Though FWS has ultimate regulatory responsibility to ensure species protection under the ESA, this novel implementation structure provides a neutral entity to assist in cross-agency coordination. An executive director and technical staff, including a chief ecologist, are responsible for program implementation and report to the GC. The executive director and staff work with official Program Advisory Committees on land, water, and science issues to implement the program's Land Plan, Water Plan, and Adaptive Management Plan (Scarlett, 2012). The executive director's office and the GC are advised by an Independent Scientific Advisory Committee on issues related to implementation of the program's Adaptive Management Plan.

At the second committee meeting held in February 2012, Gerry O'Keefe, executive director of the Headwaters Corporation, stated that the governance structure supports cross-scale and public-private coordination among multiple participants, while participants retain their individual organizational structures and identities (O'Keefe, 2012). Key challenges in shaping the governance structures and processes included deciding where to draw the negotiating boundaries and whom to include at the table. While federal and state governments and agencies have significant funding, regulatory, and decision-making responsibilities, local and nonprofit organizations directly participate in program decision-making, bringing local knowledge, values, and perspectives into the process.[2]

The driving factor that initiated action in the Platte River Basin was concern that implementing the Endangered Species Act could have significant consequences for farmers and others in the region. FWS was willing to be part of a neutral authority that brought together the stakeholders. The neutral authority obtained agreement on common goals and on monitoring to test some potential actions, thus employing adaptive management approaches.

COASTAL SYSTEMS - Great Lakes

The Great Lakes of North America are the largest body of fresh water on the planet and the largest coastal system in the lower 48 states of the United States (Figure 3-5). They have played a critical role both historically and currently in the environment, economy, and culture of the North American continent (Swackhamer, 2012). Administratively the Great Lakes are very challenging, involving two countries, eight states, two provinces, and many local governments, native peoples, and other constituencies.

The Great Lakes' enormous economic, natural resource, and social value, and the need to manage them for the need and benefit of both the United States and Canada, led to the Boundary Waters Treaty of 1909 between the two nations. This treaty has provided the foundation for more than 100 years of shared governance, which has evolved in response to various sustainability challenges to the Great Lakes.

The treaty established the International Joint Commission (IJC). IJC is comprised of three U.S. and three Canadian Commissioners who advise relevant bodies in each government on matters of national interest regarding all shared boundary waters, with a significant emphasis on the Great Lakes. In the 1960s, severe water quality problems in the Great Lakes led to the 1972 Great Lakes Water Quality Agreement, subsequently modified by protocol in 1978 and 1987, and currently being renegotiated by the two governments.

A number of other governance institutions have arisen over time. The invasion of the sea lamprey and its devastating effect on the valuable native lake trout led to the establishment of the binational Great Lakes Fisheries Commission. A subsequent group, the Great Lakes Commission, was established in 1955 by interstate legislative compact and granted Congressional consent in 1968. It is a unique governmental institution that includes the two Great Lakes Canadian provinces as formal partners. Its charge is to promote the orderly, integrated, and comprehensive development, use, and conservation of the water resources of the Great Lakes Basin.

---

[2]These organizations include, for example, irrigation companies, irrigation districts, conservancy districts, public power companies, municipalities, environmental organizations, and others.

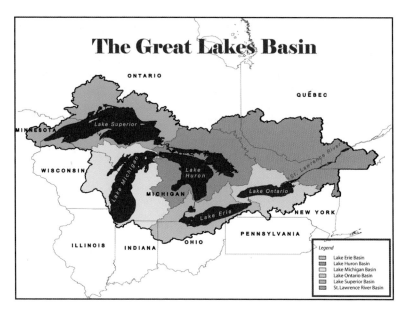

FIGURE 3-5 The Great Lakes Basin. SOURCE: Presentation by Dave Naftzger, Council of Great Lakes Governors, February 8, 2012. Image courtesy of the Great Lakes Commission.

The sustainability threats to the Great Lakes in the twentieth century include the eutrophication and ecological collapse of Lake Erie, the introduction of invasive species, and the ecological and public health threats of persistent, bioaccumulative, and toxic chemicals. We drew on the response to each of these threats to identify critical lessons learned from Great Lakes governance.

In the 1960s, Lake Erie was experiencing such an overt level of eutrophication that it was declared "dead." It became an icon for water quality problems, which were caused by excess phosphorus, whose source and actual role was hotly debated. IJC commissioned a number of prominent U.S. and Canadian scientists to model phosphorus dynamics and determine what the acceptable amount of phosphorus loading might be, in the context of what was being discharged from sewage plants. The models were the basis for discharge limits, for a decision by the U.S. EPA and Environment Canada to require primary treatment of sewage, and ultimately for the two nations' respective Clean Water Act statutes. Thus, IJC led to scientific knowledge that resulted in the Clean Water Act, sewage treatment, and discharge limits for water pollutants such as phosphorus. Without the facilitating and convening role of the IJC, it is not clear what the national trajectory of water quality management of nutrient point sources might have been.

IJC institutionalized the engagement of stakeholders around the Great Lakes Basin. The area's populace is actively invited to be engaged in and educated about Great Lakes issues and to participate in biennial meetings about priorities for the Basin. The stakeholder community, including NGOs, is generally well known, visible, and historically very active. For example, activism by this community resulted in a Canadian ban of phosphorus in detergents and the adoption of bans in the United States to reduce levels of the nutrient in water.

The existence of the Boundary Waters Treaty and the international Great Lakes Water Quality Agreement has led to other successful institutional arrangements that use science as a foundation for management and policy. For example, the International Association of Great Lakes Research consists of interdisciplinary researchers with a place-based focus. This group includes physical limnologists, fisheries biologists, aquatic biologists, aquatic chemists, ecological and human toxicologists, public health professionals, economists, sociologists, and decision scientists. It also publishes a highly cited journal. Both the U.S. and Canada have multiple federal agency research labs in the Great Lakes Basin focused on the lakes' problems. Our understanding of persistent, bioaccumulative, and toxic chemicals' behavior in the environment, accumulation in fish, and toxic effects is due to science that was facilitated by Great Lakes institutions (Swackhamer, 2012).

Another collective effort, the Great Lakes Restoration Initiative, is currently underway with significant Congressional funding. Cameron Davis, EPA, explained at the second committee meeting in February 2012 that the initiative is based on groundwork laid by an interagency task force of the federal agencies in the Basin (Davis, 2012). This task force was very effective because it already had considerable interagency cooperation and solid relationships in place.

According to Pebbles at the February 2012 second committee meeting, yet another example of institutional arrangements facilitating the use of science for decision making is the Great Lakes Fisheries Commission (Pebbles, 2012), which coordinates fisheries management in a successful partnership with the states and provinces. These partners, who hold ultimate decision-making authority, developed a Joint Strategic Plan for fisheries management based on consensus decision making informed by science, regular monitoring, and accountability. Strong relationships among the Commission, the U.S. government, and the states and provinces have aided the group's success (Stein, 2012; Figure 3-6).

The Great Lakes case reveals the important role the federal government can play in managing sustainability challenges by establishing and supporting institutions that are sustaining yet adaptable; generating scientific, social, and economic knowledge; and proactively engaging stakeholders regularly and often. The Great Lakes examples confirm that linkages in the form of federal and international agreements can enable government entities and other organizations with multiple responsibilities at multiple scales to manage, lead, and govern sustainably.

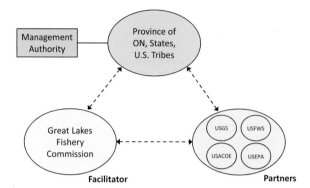

FIGURE 3-6 Science-based, Multi-Jurisdictional Fishery Management in the Laurentian Great Lakes: Exploring Federal Roles—Governance/Collaboration/Science-Decision Making. SOURCE: Presentation by Roy Stein, The Ohio State University, February 8, 2012. Image created with J. Dettmers, M. Gaden, and J. Wingfield, Great Lakes Fishery Commission.

COASTAL SYSTEMS - Pacific Northwest

The Bonneville Power Authority and the Northwest Power and Conservation Council

The nearly century-long history of dams on the Columbia River system—with their implications for energy production, water management, agriculture, forestry, recreation, and fish habitat—illustrates many crosscutting challenges in resource use, economics, and human well-being.

Exploitation of the Columbia River system for hydroelectric power and irrigation dates to the 1920s. To promote rural electrification, the Pacific Northwest Regional Planning Commission, with representatives from Idaho, Montana, Oregon, and Washington, was formed in 1934. In 1937 Congress promulgated the Bonneville Project Act, ultimately giving rise to the Bonneville Power Administration (BPA) in 1940. Electrification was seen as an economic development strategy, as a means of advancing equity between urban and rural communities, and as a path to human well-being.

As environmental awareness grew in the 1970s, attention focused on the impact of BPA's dams on fish and wildlife (Figure 3-7). In 1980 Congress passed the Pacific Northwest Electric Power Planning and Conservation Act, which required BPA to remediate damage done to fish and wildlife by its dams. A newly formed interstate compact agency, the Northwest Power and Conservation Council, was tasked with a) regional energy planning; b) fish and wildlife protection planning; and c) engagement of states, local governments, customers, federal and state fish and wildlife agencies, Indian tribes, and the public. The Act required BPA to cooperate with the Council and required BPA actions to "be consistent with" the Council's regional conservation and electric power plan—an unusual example of state authority over federal agencies, albeit an example of "soft power."

The Council's programs have protected and restored habitat for both anadromous and resident fish, launched innovative hatchery and harvest programs, and raised annual fish counts, although not to the very high levels that preceded dam construction. The Council has also been successful at integrating decision making across diverse sectors—energy, habitat restoration, irrigation, and cultural practices—and at engaging diverse publics. In its annual reports to Congress, the Council uses a range of scientific metrics of progress, another successful practice.

BPA and the Northwest Power and Conservation Council represent one of the earliest examples of shared governance with a mission and function beyond a single resource or ecological dimension. Key lessons learned from the management of the Columbia River watershed and its hydroelectric resources include the potential effectiveness of devolving authority from the federal government to regional players, the ability of these regional players to convene voices from diverse sectors, and the benefits of exercising "soft power" rather than rigid authority.

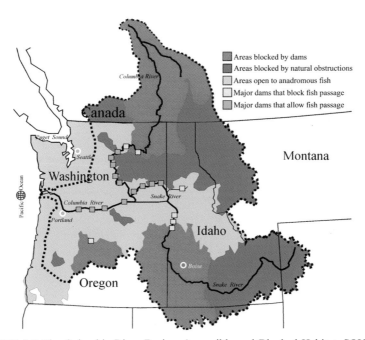

FIGURE 3-7 The Columbia River Basin—Accessible and Blocked Habitat. SOURCE: Presentation by Phil Rockefeller, Northwest Power and Conservation Council, February 7, 2012.

Puget Sound Partnership

A collaborative initiative with very broad participation, the Puget Sound Partnership (PSP) is an agency established under an executive order of the Governor of Washington. PSP is a state agency and as a result has a line item budget from the state legislature. The partnership collaborates with EPA and is deeply involved with its Action Plan for Puget Sound and the Northwest Power and Conservation Council. PSP includes citizens, tribes, governments, business, and scientists working together with the objective of protecting and restoring the Sound. The primary focus of the partnership, as seen in its goals and objectives, appears to be the ecological health of the Sound and only indirectly the economic, social, and health dimensions of a sustainable region. While those involved in the partnership express a clear understanding of the interdependence of the ecology of Puget Sound and the health and economic well-being of the region's communities, these considerations are not yet reflected in the partnership's goals.

PSP is challenged in that it holds no authority to implement its policies, findings, or agreements; this authority instead resides with its member organizations (O'Keefe, 2012). The partnership's lack of authority has led to problems achieving stated goals. In 2012, PSP's draft State of the Sound report acknowledged in a review of available data that to date, progress has not been sufficient to meet 2020 recovery targets (PSP, 2012). Of the 21 indicators, PSP found that two showed clear progress, five showed mixed results, seven demonstrated no progress, and seven were considered incomplete because there were no data or because the targets had not been adopted or were still in development (PSP, 2012). A report by the State of Washington Joint Legislative Audit and Review Committee (JLARC, 2013) stated that in a previous review of the Partnership, the 2008 Action Agenda lacked critical accountability tools, which are fundamental to determining whether the funding spent on clean-up efforts is restoring Puget Sound. A 2013 follow up report noted that "while the 2012 Action Agenda has taken steps to improve accountability, it continues to have shortcomings in three key areas: linkages, prioritization, and monitoring" (JLARC, 2013). Specifically, the review notes that the 2012 Action Agenda does not link actions to the amount of progress they will make toward long-term restoration goals. Additionally, a significant amount of funding has been spent on programs that have not been prioritized; although the PSP recommends that all effective ongoing programs be maintained, "it has not identified which ones are effective" (JLARC, 2013). Sufficient mechanisms are not in place to monitor actions in order to understand which ones are working (JLARC, 2013).

In addition, PSP has inadequate representation and participation by land-use authorities. Because land use is such a critical factor affecting ecological, economic, and social health, involving local land-use authorities presents a significant opportunity for collaboration. A 2012 report from the Washington State Academy of Sciences (WSAS) reiterated the importance of land use in the discussion and the need to develop related measures of progress, stating that "many of the important environmental changes caused by humans…result from land use." WSAS recommended that an indicator set that adequately characterizes the "condition of Puget Sound needs to include indicators that represent the extent of each habitat type and other measures of marine landscape pattern and structure" (WSAS, 2012).

OPPORTUNITIES FOR ADDRESSING SUSTAINABILITY LINKAGES: LESSONS LEARNED

While the examples reviewed by the committee varied in their details, they provide some common themes and insights on ways to enhance sustainability linkages in federal decision making. These lessons address issues of governance, decision-making processes, and science.

1. Iterative improvements: Enhancing governance linkages to address sustainability does not need to occur through disruptive change, and in fact is generally more successful through iterative change with incremental steps. Positive benefits from incremental changes, if documented properly and articulated, can lead to broader and ultimately comprehensive change without the paralysis that can sometimes be associated with proposals for disruptive change. Although the challenges of integrating decision making across domains and among agencies at various levels of government are significant, the examples we examined demonstrate innovations to integrate decisions that vary from modest and iterative steps to more substantial governance redesign. The former initiatives should not be discounted merely because they are smaller, for they can often lead to significant accomplishments. For example, DMG in the Mojave, a pioneering effort to better coordinate information and enhance dialogue among multiple federal, state, and local agencies, did not involve any agency restructuring or any enhanced or new decisionmaking authorities. The primary purpose was to improve dialogue among agencies with responsibilities within a single geographic area, within existing structures and authorities. DMG is an important partnership that built trust and enabled the successful formation and function of the California REAT and the development of the DRECP. Similarly, the success of the effort in Platte River Basin arose largely from the enforcement of an existing regulation implementing the ESA. FWS was willing to be part of a neutral authority that brought together the stakeholders; this neutral authority obtained agreement on common goals and on monitoring to test some potential actions, thus employing adaptive management approaches.

2. Multiple levels of government: Several of the fact-finding examples illustrate that change agents engaged in innovations to strengthen sustainability linkages in decisionmaking arise at all levels of government. Many examples the committee reviewed were initiated at the local or field level (bottom-up), as multiple federal, state, and local agencies strived to grapple with linked issues. Examples drawn from literature include the Sonoita Valley Planning Partnership in the Cienegas Watershed south of Tucson and the Penobscot River Restoration

Project in the northeast. In both of these cases, federal agencies are critical partners, but the motivation for action started at a grassroots level. At the same time, the committee also reviewed examples in which national and even international agencies or forums initiated collaborative, interconnecting actions. In some instances, federal laws may have prompted the need for regulatory compliance, which in turn motivated creative local action to integrate decisions across interconnected agencies and issues. For example, EPA regulations regarding CSO control triggered a need for Philadelphia to bring its infrastructure into compliance. The high costs of traditional compliance tools prompted the city to explore nonstructural alternatives, including extensive expansion of permeable surfacing. The extent of transformation needed to meet the stormwater regulations motivated the city to work across multiple agencies and examine co-benefits in other domains. In the Great Lakes, a longstanding international treaty provided a forum that helped spark action at international, state, and local levels to address water quality needs in the region. Even within the Great Lakes context, however, some actions have been largely locally motivated. In addition, it is important to note that local and statewide policy efforts can also play a major role in sustainability efforts.

   3. Network governance: Several of the examples examined by the committee illustrate the emergence of network governance models to enhance coordinated decision making and address sustainability linkages. The concept of network governance surfaced first in the private sector as corporations working together on joint projects developed horizontal, or shared, governance structures. Applications of network governance models are also appearing in the public sector (Goldsmith and Eggers, 2004). Such governance is characterized by a polycentric (multi-participant, multi-agency) approach, often operating with self-constituted decision rules determined through negotiation and cooperative agreements among participants. These governance structures provide a fabric for cross-domain, interagency, and public-private coordination without restructuring existing agencies or reallocating statutory authorities. Examples reviewed by the committee include the Platte River Recovery Governance Committee structure and decision process, the California Renewable Energy Action Team, and the Puget Sound Partnership. Numerous other examples vary along a continuum from loosely knit confederations to congressionally authorized, formal, interagency coordinating structures.

   4. Stakeholders at the table: The importance of having a full and adequate representation of all affected stakeholders in partnerships and other forms of collaborative governance structures is well understood. In addition, the necessity of reaching consensus on goals, roles, responsibilities, and accountability is well documented (NRC, 2011b). Also, the committee recognizes the importance of having an agreed-upon process for decisionmaking that allows for a balanced evaluation of different development scenarios under the sustainability lens. Examination of the more successful collaborative governance efforts—the Bonneville Power Authority, the Platte River Recovery Program, and the Mojave DMG—highlighted the importance of full participation by parties that represented the key drivers affecting sustainable outcomes. In each of these cases, the organizations responsible for activities with the greatest influence on regional sustainability (Bonneville Power on the Columbia River, the states of Wyoming, Colorado, and Nebraska on the Platte River, and federal land agencies in the Mojave) were central to the collaborative effort. Their support of consensus goals and implementation was critical. Conversely, within Puget Sound, the parties responsible for land use, a very important driver affecting the economy, human well-being, and environmental quality, are not included in the Puget Sound Partnership.[3] In this case, a more complete analysis of the problems in the region during the framing process and the identification of all relevant players, as described in the committee's proposed decision framework (see Chapter 4), would have enhanced the effectiveness of the partnership. Land-use authorities in some other locales do participate in collaborative governance efforts, however. One example in the literature is the Boston Harbor Island National Recreation Area, a network governance structure that plans and manages a mosaic of state, local, and nonprofit lands; the group includes federal, state, and local agencies with land management responsibilities (Boston Harbor Islands Partnership Charter, 2006).

5. Mutual learning, interdisciplinary partnering, and trust: A strong science base with open dialogue and partnering among scientists, decision makers, and stakeholders is a hallmark of successful sustainability efforts. Joint research efforts on the Great Lakes that involve academic and government laboratories in both the United States and Canada led to federal standards in both countries to improve water quality and to sustain commercial and recreational fisheries. The power of long-term research in maintaining biodiversity was clearly evident in partnerships between the ASU-led CAP LTER project (supported by NSF and others) and local officials and NGOs in Phoenix. Similarly, open sharing of data, research findings, and ecosystem maps among stakeholders was critical to finding a path forward for the DMG of the Mojave. Research on connections between violent crime and vacant lots led to partnerships between NGOs and the City of Philadelphia to promote "greening" efforts. Partnerships between fisheries experts from government laboratories and academia, as well as between decision makers and other stakeholders, played a key role in protecting salmon at the Bonneville dam. These examples demonstrate that interdisciplinary, place-based research is often a vital part of addressing sustainability linkages. For example, increased temperatures disproportionally affect vulnerable populations in urban area heat islands. Predictive models of climate change indicate that temperatures will increase in Phoenix by mid-century to potentially dangerous levels (heat waves with durations of up to 52 days with temperatures above 122 degrees F). More broadly, changes in climate will im-

---

[3]To a significant extent, achievement of the goals of the Puget Sound Partnership relative to salmon habitat and population, water quality, regional transportation systems, climate change adaptation, and many others could not be achieved without agreement on land use.

pact a range of disease vectors and will require planning and preparation to protect vulnerable populations.

6. Adaptive management: Many sustainability challenges that involve interconnections among domains—for example, transportation, energy, water, health, and species protection—are complex and dynamic, resulting in uncertainties about current interactions among variables, cause-effect relationships, and projected future conditions. As a consequence of this complexity and uncertainty, participants in a number of the examples we examined are using an adaptive management approach, as described in the decision framework (see Chapter 4). Adaptive management enables participants to set goals, undertake actions, monitor the effects of those actions on outcomes, and, most importantly, make adjustments as needed. In the Platte River Recovery Implementation Plan, adaptive management processes help managers address uncertainties regarding what water management regimes will best meet the needs of endangered species while, at the same time, sustaining sufficient water for agriculture, energy, and other uses. In addition, this approach was attributed with helping the plan's participants transcend scientific disagreements regarding the amount and timing of water flows necessary for species protection. To date, application of adaptive management has had limited implementation success in changing management actions based on experimentation and monitoring (Murray and Marmorek, 2004; Kimberly et al., 2006). However, recent assessments of adaptive management indicate that effectiveness in using the approach can be enhanced by: 1) starting with a simple plan and adding complexity over time; and 2) engaging researchers at all stages of the process (Kimberly et al., 2006).

7. Creative approaches to problem-solving can add value and provide multiple benefits or co-benefits to participants: Innovative thinking that crosses domains can result in sustainability solutions that increase efficiency and cost-effectiveness and that create win-win scenarios. For example, Philadelphia dealt with the treatment of storm water by providing a cost-effective solution that resulted in multiple benefits. Working with decision makers at EPA, the city was able to reduce the need for additional costly infrastructure improvements by utilizing a variety of approaches to reduce the volume of water run-off and to take advantage of natural ameliorative processes in soils and subsurface environments. At the same time, this green infrastructure approach was shown by the city to achieve co-benefits for outdoor recreation, public health, education, and the local economy.

8. Communications: Sustainability solutions need to be communicated in a way that clearly identifies both the costs and benefits of action and inaction. For example, Philadelphia got an enormous boost for its approach when sustainability became a plank in the mayoral campaign. An effective communications strategy is important not only at the outset to engage major and important constituencies, but also throughout the process in keeping key stakeholders and the public generally aware of the progress being made and the work that still needs to be done. Effective communications and stakeholder participation also promotes transparency and accountability.

REFERENCES

Alberti, M., J. A. Hepinstall, S. E. Coe, R. Coburn, M. Russo and Y. Jiang. Modeling Urban Patterns and Landscape Change in Central Puget Sound. Online. Available at http://www.isprs.org/proceedings/XXXVI/8-W27/Alberti.pdf. Accessed August 31, 2012.

American Public Transportation Association. 2011. Sep-tainable: The Route to Regional Sustainability. Washington, DC: American Public Transportation Association.

American Public Transportation Association. 2012. Sep-tainable: Going Beyond Green. Washington, DC: American Public Transportation Association.

ASU (Arizona State University) Morrison Institute for Public Policy. 2011. Watering the Sun Corridor: Managing Choices in Arizona's Megapolitan Area. Online. Available at http://morrisoninstitute.asu.edu/publications-reports/2011-watering-the-sun-corridor-managing-choices-in-arizonas-megapolitan-area. Accessed September 28, 2012.

Becker, A., S. Inoue, M. Fischer, and B. Schwegleret. 2012. Climate change impacts on international seaports: knowledge perceptions, and planning efforts among port administrators. *Climatic Change* 110:5-29.

Branas. C. C. 2012. Public Health Issues and Sustainability. Presentation to National Research Council Committee on Sustainability Linkages in the Federal Government. June 12, 2012.

Branas, C. C., R. A. Cheney, J. M. MacDonald, V. W. Tam, T. D. Jackson, and T. R. Ten Have. 2011. A difference-in-differences analysis of health, safety, and greening vacant urban space. *American Journal of Epidemiology* 174(11):1296-1306.

Brunner, P. H. 2007. Materials flow analysis: Reshaping urban metabolism. *Journal of Industrial Ecology* 11(2):11-13.

Byron, C., D. Bengtson, B. Costa-Pierce, and J. Calanni. 2011. Integrating science into management: Ecological carrying capacity of bivalve shellfish aquaculture. *Marine Policy* 35:363-370.

California Department of Fish and Game, California Energy Commission, California Bureau of Land Management, and the U.S. Fish and Wildlife Service. 2008. Memorandum of Understanding between the California Department of Fish and Game, the California Energy Commission, the Bureau of Land Management, and the U.S. Fish and Wildlife Service regarding the Establishment of the California Renewable Energy Permit Team. Online. Available at http://www.blm.gov/pgdata/etc/media lib/blm/ca/pdf/pa/energy.Par.76169.File.dat/RenewableEnergyMOU-CDFG-CEC-BLM-USFWS-Nov08.pdf. Accessed August 31, 2012.

Center for the Future of Arizona. 2009. The Arizona We Want. Phoenix, AZ: Center for the Future of Arizona.

Ciborowski, J. J. H., G. J. Niemi, V. J. Brady, S. E. Doka, L. B. Johnson, J. R. Keough, S. D. Mackey, and D. G. Uzarski. 2009. Ecosystem responses to regulation-based water level changes in the Upper Great Lakes. *White paper*:1-56.

City of Philadelphia, Mayor's Office of Sustainability. 2009. Greenworks Philadelphia. Online. Available at http://www.phila.gov/green/greenworks/pdf/Greenworks_On linePDF_FINAL.pdf Accessed March 4, 2013.

Crockett, C. 2012. Water Resources and Sustainability in Philadelphia. Presentation to the National Research Council's Committee on Sustainability Linkages in the Federal Government, Fourth Meeting. June 12, 2012.

Cuo, L., T. K. Beyene, N. Voisin, F. Su, D. P. Lettenmaier, M. Albert, and J. E. Richey. 2010. Effects of mid-twenty-first century climate and land cover change on the hydrology of the Puget Sound basin, Washington. *Hydrological Processes* 25:1729-1753.

Davis, C. 2012. Presentation to the National Research Council's Committee on Sustainability Linkages in the Federal Government, Second Meeting. February 8, 2012.

Delaware Valley Green Building Council. 2012. Sustainable Water Strategies in Philadelphia: Toward Green Building Practices that Conserve, Reuse, and Manage Water. Philadelphia, PA: Delaware Valley Green Building Council.

Dinse, K., J. Read, and D. Scavia. 2009. Preparing for Climate Change in the Great Lakes Region. Ann Arbor, MI: Michigan Sea Grant.

DRECP (Desert Renewable Energy Conservation Plan). 2012. Online. Available at http://www.drecp.org/about/index.html. Accessed February 28, 2013.

DRECP Independent Science Advisors. 2010. Recommendations of Independent Science Advisors for the California Desert Renewable Energy Conservation Plan. Corvallis, OR: Conservation Biology Institute.

Delaware Valley Green Building Council. 2012. Sustainable Water Strategies in Philadelphia: Toward Green Building Practices that Conserve, Reuse, and Manage Water. Philadelphia, PA: Delaware Valley Green Building Council.

DMG (Desert Managers Group). 2011. Overview, Strategic Pan, Charter and MOU. Barstow, CA: DMG.

DMG. December 10, 2010. NPS Partnership Case Study. Barstow, CA: DMG.

DOI (U.S. Department of Interior). 2004. The Platte River Channel: History and Restoration. Denver, CO: Technical Service Center.

DOI and State of California. 2009. Memorandum of Understanding between Department of Interior and State of California. Online. Available at http://www.doi.gov/documents/CAMOUsigned.pdf. Accessed August 31, 2012.

Elsner, M. M., L. Cuo, N. Voisin, J. S. Deems, A. F. Hamlet, J. A. Vano, K. E. B. Mickelson, S. Lee, and D. P. Lettenmaier. 2010. Implications of 21st century climate change for the hydrology of Washington State. *Climatic Change* 102:225-260.

EPA (U.S. Environmental Protection Agency) and Environment Canada. 2011. The Great Lakes: Environmental Atlas and Resource Book. Chicago, IL: Great Lakes National Program Office.

EPA. Partnership for Sustainable Communities: An Interagency Partnership of HUD, DOT, & EPA. Online. Available at http://www.epa.gov/dced/partnership/#livabili typrinciples. Accessed September 4, 2012.

EPA. 2011. Urban Waters Federal Partnership: Vision, Mission & Principles. Online. Available at http://www.urbanwaters.gov/pdf/urbanwaters-visionv2012.pdf. Accessed September 4, 2012.

Feely, R. A., S. R. Alin, J. Newton, C. L. Sabine, M. Warner, A. Devol, C. Krembs, and C. Maloy. 2010. The combined effects of ocean acidification, mixing, and respiration on pH and carbonate saturation in an urbanized estuary. *Estuarine, Coastal and Shelf Science* 88:442-449.

Fraser, D. A., J. K. Gaydos, E. Karlsen, and M. S. Rylko. 2006. Collaborative science, policy development and program implementation in the transboundary Georgia Basin/Puget Sound ecosystem. *Environmental Monitoring and Assessment* 113:49-69.

Freeman, D. 2010. Implementing the Endangered Species Act on the Platte Basin Water Commons. Colorado: University of Colorado Press.

Freeman, D. 2012. Presentation to the National Research Council's Committee on Sustainability Linkages in the Federal Government, Third Meeting. April 12, 2012.

Glicksman, R. L., C. O'Neill, Y. Huang, W. L. Andreen, R. K. Craig, V. Flatt, W. Funk, D. Goble, A. Kaswan, and R. R. M. Verchick. 2011. Climate Change and the Puget Sound: Building the Legal Framework for Adaptation. Center for Progressive Reform White Paper No. 1108. Online. Available at http://www.progressivereform.org/articles/Puget_Sound_Adaptation_1108.pdf. Accessed November 9, 2012.

Good, T. P., J. A. June, M. A. Etnier, and G. Broadhurst. 2010. Derelict fishing nets in Puget Sound and the Northwest Straits: Patterns and threats to marine fauna. *Marine Pollution Bulletin* 60:39-50.

Grimm, N. B., and C. L. Redman. 2004. Approaches to the study of urban ecosystems: the case of Central Arizona—Phoenix. *Urban Ecosystems* 7:199-213.

Hall, N. D. 2006. Toward a new horizontal federalism: interstate water management in the Great Lakes region. *University of Colorado Law Review* 77:405-456.

Harlan, S. L. 2012. Environmental Injustice in a Desert City: Green Spaces, Heat and Social Inequality. Prepared for the Arid LID Conference, Tucson, Arizona. Online. Available at http://www.aridlid.org/wp-content/uploads/2012/06/Harlan.pdf. Accessed February 26, 2013.

Harlan, S. L., A. J. Brazel, L. Prashad, W. L. Stefanov, and L. Larsen. 2006. Neighborhood microclimates and vulnerability to heat stress. *Social Science & Medicine* 63:2847-2863.

Harlan, S. L., J. H. Declet-Barreto, W. L. Stefanov, and D. Petitti. 2013. Neighborhood Effects on Heat Deaths: Social and Environmental Determinants of Vulnerability in Maricopa County, Arizona. *Environmental Health Perspectives* 121(2):197-204.

Hartig, J. H., M. A. Zarull, J. J. H. Ciborowski, J. E. Gannon, E. Wilke, G. Norwood, and A. N. Vincent. 2009. Long-term ecosystem monitoring and assessment of the Detroit River and Western Lake Erie. *Environmental Monitoring and Assessment* 158:87-104.

Hebert, C. E., D. V. C. Weseloh, A. Idrissi, M. T. Arts, and E. Roseman. 2009. Diets of aquatic birds reflect changes in the Lake Huron ecosystem. *Aquatic Ecosystem Health & Management* 12(1):37-44.

Hepinstall-Cymerman, J., S. Coe, and L. R. Hutyra. 2011. Urban growth patterns and growth management boundaries in the Central Puget Sound, Washington, 1986-2007. *Urban Ecosystems*. DOI 10.1007/s11252-011-0206-3.

Hess, J. J., J. N. Malilay, and A. J. Parkinson. 2008. Climate change: The importance of place. *American Journal of Preventive Medicine* 35(5):468-478.

Hildebrand, L. P., V. Pebbles, and D. A. Fraser. 2002. Cooperative ecosystem management across the Canada-US border: Approaches and experiences of transboundary programs in the Gulf of Maine, Great Lakes and Georgia Basin/Puget Sound. *Ocean & Coastal Management* 45:421-457.

Iceland, C., C. Hanson, and C. Lewis. 2008. Identifying Important Ecosystem Goods & Services in Puget Sound. Draft summary of interviews and research for the Puget Sound Partnership. Online. Available at http://www.psp.wa.gov/downloads/AA2008/ecosystem_services_analysis.pdf. Accessed August 31, 2012.

JLARC (State of Washington Joint Legislative Audit and Review Committee). 2013. PSP'S 2012 Action Agenda Update: Revised Approach Continues to Lack Key Accountability Tools Envisioned in Statute. Briefing Report. Online. Available at http://www.leg.wa.gov/JLARC/AuditAndStudyReports/2013/Documents/PugetSoundPartnershipFollowUpReport.pdf. Accessed March 26, 2013.

Kennedy, C., J. Cuddihy, and J. Engel-Yan. 2007. Research and analysis: The changing metabolism of cities. *Journal of Industrial Ecology* 11(2):43-59.

Kimberly, J., R. Morghan, R. L Sheley, and T. J. Svejcar. 2006. Successful Adaptive Management—The Integration of Research and Management. *Rangeland Ecology and Management* 59(2):216-219.

Kois, M. A. 2012. Positive outlook for urban greening of vacant lots. *Living Architecture Monitor* 14(1):24-28.

Krantzberg G. 2009. Renegotiating the Great Lakes water quality agreement: The process for a sustainable outcome. *Sustainability* 2009(1):254-267.

Kwartin, R., S. Alexander, M. Anderson, D. Clark, J. Collins, C. Lamson, G. Martin, R. Mayfield, L. McAlpine, D. Moreno, J. Patterson, C. Schultz, and E. Stiever. 2012. Solar Energy Development on Department of Defense Installations in the Mojave and Colorado Deserts. Washington, DC: ICF International.

Layzer, J. A., and S. B. Stern. 2010. What Works and Why?: Evaluating the Effectiveness of Cities' Sustainability Initiatives. Prepared for the American Political Science Association Meeting, September 2-5, 2010, Washington, D.C.

Leschine, T. M. 2010. Human dimensions of nearshore restoration and shoreline armoring with application to Puget Sound. Pp. 103-114 in Puget Sound Shorelines and the Impacts of Armoring—Proceedings of a State of the Science Workshop, May 2009, H. Shipman, M. N. Dethier, G. Gelfenbaum, K. L. Fresh, and R. S. Dinicola, eds. U.S. Geological Survey Scientific Investigations Report 2010-5254.

Liu, J., T. Dietz, S. R. Carpenter, M. Alberti, C. Folke, E. Moran, A. N. Pell, P. Deadman, T. Kratz, J. Lubchenco, E. Ostrom, Z. Ouyang, W. Provencher, C. L. Redman, S. H. Schneider, and W. W. Taylor. 2007. Complexity of coupled human and natural systems. *Science* 317:1513-1516.

McCue, K. 2012. Conservation of Threatened Species and Habitats. Presentation to the National Research Council's Committee on Sustainability Linkages in the Federal Government, Fourth Meeting. June 11, 2012.

McGranahan, G., and D. Satterthwaite. 2003. Urban centers: An assessment of sustainability. *Annual Review of Environment and Resources* 28:243-74.

McKay, N. 1991. Environmental management of the Puget Sound. *Marine Pollution Bulletin* 23:509-512.

Mohapatra, S. P., and A. Mitchell. 2009. Groundwater demand management in the Great Lakes Basin—Directions for new policies. *Water Resources Management* 23:457-475.

Morley, S. A., and J. R. Karr. 2002. Assessing and restoring the health of urban streams in the Puget Sound Basin. *Conservation Biology* 16(6):1498-1509.

Morrison Institute for Public Policy. 2011. Watering the Sun Corridor: Managing Choices in Arizona's Megapolitan Area. Tempe, AZ: Arizona State University.

Mortsch, L. D. 1998. Assessing the impact of climate change on the Great Lakes shorelines and wetlands. *Climatic Change* 40:391-416.

Murray, C., and D. R. Marmorek. 2004. Adaptive Management: A Spoonful of Rigour Helps the Uncertainty Go Down. Presentation at the 16th International Annual Meeting of the Society for Ecological Restoration. Victoria, British Columbia, August 23-27, 2004.

Naftzger, D. 2012. Management and Governance in the Great Lakes Region. Presentation to the National Research Council's Committee on Sustainability Linkages in the Federal Government, Second Meeting. February 8, 2012.

Neukrug, H. M. 2011. Sustainability Linkages in the Federal Government. Presentation to the National Research Council's Committee on Sustainability Linkages in the Federal Government, First Meeting. September 20, 2011.

NPS (National Park Service). 2003. Partnerships. Online. Available at http://www.nps. gov/partnerships/ca_dmg.htm. Accessed March 11, 2013.

NRC (National Research Council). 2010. Pathways to Urban Sustainability: Research and Development on Urban Systems. Washington, DC: National Academies Press.

NRC. 2011a. Pathways to Urban Sustainability: Lessons from the Atlanta Metropolitan Region: Summary of a Workshop. Washington, DC: National Academies Press.

NRC. 2011b. Sustainability and the U.S. EPA. Washington, DC: National Academies Press.

Nutter, M. A. Greenworks Philadelphia. Philadelphia, PA: The City of Philadelphia. Online. Available at http://www.phila.gov/green/greenworks/PDFs/GreenworksPl an002.pdf. Accessed October 1, 2012.

Office of the Governor. 2008. Executive Order S-14-08. Online. Available at http://gov. ca.gov/news.php?id=11072. Accessed August 31, 2012.

Oki, T., and S. Kanae. 2006. Global Hydrological Cycles and World Water Resources. Science 313(5790):1068-1072.

O'Keefe, G. 2012. Presentation to the National Research Council Committee on Sustainability Linkages in the Federal Government, Second Meeting. February 7, 2012.

Ormerod, S. J., M. Dobson, A. G. Hildrew, and C. R. Townsend. 2010. Multiple stressors in freshwater ecosystems. *Freshwater Biology* 55(Suppl. 1):1-4.

Overpeck, J. T., and B. Udall. 2010. Dry times ahead. *Science* 328:1642-1643.

Pebbles, V. Presentation to the National Research Council's Committee on Sustainability Linkages in the Federal Government, Second Meeting. February 8, 2012.

PennPraxis. 2010. Green2015: An Action Plan for the First 500 Acres. Prepared for Philadelphia Parks and Recreation. Philadelphia, PA: University of Pennsylvania.

Petitti, K. 2012. Presentation to the National Research Council's Committee on Sustainability Linkages in the Federal Government, Fourth Meeting. June 11, 2012.

Phama, H. M., Y. Yamaguchia, and T. Q. Buib. 2011. A case study on the relation between city planning and urban growth using remote sensing and spatial metrics. *Landscape and Urban Planning* 100:223-230.

PWD (Philadelphia Water Department). 2011. Amended Green City Clean Waters: The City of Philadelphia's Program for Combined Sewer Overflow Control. June 1, 2011. Online. Available at http://www.phillywatersheds.org/doc/GCCW_Amended June2011_LOWRES-web.pdf. Accessed August 31, 2012.

Platte River Recovery Implementation Program. 2007. Governance Committee Meeting Minutes. Online. Available at http://www.platteriverprogram.org/PubsAndData/Pr ogramLibrary/2007%20August%20GC%20Minutes.pdf. Accessed October 1, 2012.

Platte River Recovery Implementation Program. 2010. Bi-Annual Report 2009-2010. Kearney, NE: Headwaters Corporation.

PSP (Puget Sound Partnership). 2012. 2012 State of the Sound. Online. Available at http://www.psp.wa.gov/sos.php. Accessed March 26, 2013.

Quay, R. 2004. Bridging the gap between ecological research and land use policy: The North Sonoran Collaboration. *Urban Ecosystems* 7:283-294.

Quay, R. 2010. Anticipatory governance: A tool for climate change adaptation. *Journal of the American Planning Association* 76(4):496-511.

Quay, R. 2012. Presentation to the National Research Council's Committee on Sustainability Linkages in the Federal Government, Fourth Meeting. June 11, 2012.

Rockefeller, P. 2012. Presentation to the National Research Council's Committee on Sustainability Linkages in the Federal Government, Second Meeting. February 8, 2012.

Rosenberg, E. A., P. W. Keys, D. B. Booth, D. Hartley, J. Burkey, A. C. Steinemann, and D. P. Lettenmaier. 2010. Precipitation extremes and the impacts of climate change on stormwater infrastructure in Washington State. *Climatic Change* 102:319-349.

Roy, E. D., J. F. Martin, E. G. Irwin, J. D. Conroy, and D. A. Culver. 2010. Transient social-ecological stability: The effects of invasive species and ecosystem restoration on nutrient management compromise in Lake Erie. *Ecology and Society* 15(1):20.

SEPTA (Southeastern Pennsylvania Transportation Authority). 2011. Sep-tainable: The Route to Regional Sustainability. Online. Available at http://www.septa.org/sustain/pdf/septainable11.pdf. Accessed March 4, 2013.

Scarlett, L. 2010. Green, Clean and Dollar Smart Ecosystem Restoration in Cities and Countryside. Washington, DC: Environmental Defense Fund.

Scarlett, L. 2012. Managing Water: Governance Innovations to Enhance Coordination. Issue Brief 12-04. Resources for the Future. Online. Available at: http://www.rff.org/RFF/Documents/RFF-IB-12-04.pdf. Accessed March 11, 2013.

Schauman, S., and S. Salisbury. 1998. Restoring nature in the city: Puget Sound experiences. *Landscape and Urban Planning* 42:287-295.

Scofield, R. 2012. Presentation to the National Research Council's Committee on Sustainability Linkages in the Federal Government, Third Meeting. April 11, 2012.

Senate Energy, Utilities and Communications Committee. 2011. SBX1 2 - Simitian. Online. Available at http://www.leginfo.ca.gov/pub/11-12/bill/sen/sb_0001-0050/sbx1_2_cfa_20110214_141136_sen_comm.html. Accessed September 28, 2012.

Skaggs, R., T. C. Janetos, K. A. Hibbard, and J. S. Rice. 2012. Climate and Energy-Water-Land System Interactions: Technical Report to the U.S. Department of Energy in Support of the National Climate Assessment. Richland, WA: Pacific Northwest National Laboratory.

Smith, C. B. 2011. Adaptive management on the central Platte River—Science, engineering, and decision analysis to assist in the recovery of four species. *Journal of Environmental Management* 92(5):1414-1419.

Sproule-Jones, M. 2008. Transboundaries of Environmental Governance on the Great Lakes. Online. Available at http://www.indiana.edu/~workshop/colloquia/materials/papers/sproule-jones_paper.pdf. Accessed August 30, 2012.

Stein, R. 2012. Presentation to the National Research Council's Committee on Sustainability Linkages in the Federal Government, Second Meeting. February 8, 2012.

Swackhamer, D. L. 2012. Water quality and sustainability in the Great Lakes: Persistent organic pollutants. In Comprehensive Water Quality and Purification. Volume 4: Water Quality and Its Sustainability, J. Schnoor, ed. Amsterdam, Netherlands: Elsevier Press.

Thormodsgard, J. M. 2009. Greater Platte River Basins—Science to Sustain Ecosystems and Communities. *U.S. Geological Survey Fact Sheet 2009-3097.*

Turner II, B. L., A. J. Janetos, and P. H. Verburg. The Architecture of Land Systems: A Novel Strategy for Global Environmental Change and Sustainability Science and Policy.

University of Nebraska-Lincoln Office of Research. 2008. Sustainability in a Time of Climate Change: Developing an Intensive Research Framework for the Platte River Basin and the High Plains. Lincoln, NE: University of Nebraska-Lincoln.

Vano, J. A., N. Voisin, L. Cuo, A. F. Hamlet, M. M. Elsner, R. N. Palmer, A. Polebitski, and D. P. Lettenmaier. 2010. Climate change impacts on water management in the Puget Sound region, Washington State, USA. *Climactic Change* 102:261-286.

WSAS (Washington State Academy of Sciences). 2012. Sound Indicators: A Review for the Puget Sound Partnership. Online. Available at http://www.washacad.org/about/files/WSAS_Sound_Indicators_wv1.pdf. Accessed March 26, 2013.

White, D. D., A. Wutich, K. L. Larson, P. Gober, T. Lant and C. Senneville. 2010. Credibility, salience, and legitimacy of boundary objects: water managers' assessment of a simulation model in an immersive decision theater. *Science and Public Policy* 37(3):219-232.

White, D. D., E. A. Corley., and M. S. White. 2008. Water managers' perceptions of the science-policy interface in Phoenix, Arizona: Implications for an emerging boundary organization. *Society and Natural Resources* 21:230-243.

Wilbanks, T., S. Fernandez, G. Backus, P. Garcia, K. Jonietz, P. Kirshen, M. Savonis, B. Solecki, L. Toole, M. Allen, R. Bierbaum, T. Brown, N. Brune, J. Buizer, J. Fu, O. Omitaomu, L. Scarlett, M. Susman, E. Vugrin, and R. Zimmerman. 2012. Climate Change and Infrastructure, Urban Systems, and Vulnerabilities: Technical Report for the U.S. Department of Energy in Support of the National Climate Assessment. Oak Ridge, TN: Oak Ridge National Laboratory.

World Bank. 2011. Guide to Climate Change Adaptation in Cities. Washington, DC: The World Bank.

Xuemei, B. 2007. Industrial ecology and the global impacts of cities: Editorial. *Journal of Industrial Ecology* 11(2):1-6.

Yang, Z., and T. Khangaonkar. 2010. Multi-scale modeling of Puget Sound using an unstructured-grid coastal ocean model: from tide flats to estuaries and coastal waters. *Ocean Dynamics* 60:1621-1637.

Chapter 4

Development of a Decision Framework

THE NEED FOR AND VALUE OF A DECISION FRAMEWORK

The preceding chapters identified the need for a consistent decision framework that can be used to strengthen sustainability linkages. Drawing from a number of the fact-finding examples and the literature, the committee identified the common elements of an effective decision framework, which form the basis for the framework presented here.

Decision frameworks provide a way to facilitate and enhance decision making by providing conceptual structures and principles for integrating the economic, social, ecological, and legal/institutional dimensions of decisions. Their application can result in consistent and effective results. Decision frameworks refer to principles, processes, and practices to proceed from information and desires to choices that inform actions and outcomes (Lockie and Rockloff, 2005).

While decision frameworks vary in design and purpose, they generally have common elements that include:

- Problem identification and formulation,
- Identification of clear goals,
- Illumination of key questions that help decision participants scope problems and management options,
- Processes for knowledge-building (including scientific, technical, experiential, and cultural knowledge) and application of appropriate analytical tools to assess actions, options, trade-offs, risks, and uncertainties,
- Connection of authorities tasked with making decisions to outcomes associated with those decisions.

In addition to these common elements, decision frameworks generally provide transparency about goals, information, and decision processes; inclusiveness of relevant participants; and structures or processes to adapt decisions over time in response to new goals, changing circumstances, or new knowledge.

In this chapter, the principles that form the basis for the decision framework the committee recommends are first articulated, followed by the framework itself. Recommendations concerning its implementation and use are also presented.

PRINCIPLES

The decision framework described in this chapter was developed to be:

- Flexible and scalable to a wide range of complex sustainability issues
- Based on the broad and diverse literature and practice of effectively and widely used frameworks
- Inclusive of the major elements of such frameworks

As illustrated in the examples addressed in this report, any framework must be flexible enough that it can be applied to a broad range of sustainability linkage challenges. Consequently, for the decision framework presented here to be broadly useful, it must be sufficiently flexible to be adapted to a wide range of applications. As also illustrated in this report, sustainability linkage applications vary both temporally and geographically. Consequently, the decision framework must also be scalable.

A broad and diverse literature and significant practical experience with decision frameworks exist (see Box 4-1). This literature and experience provide the foundation for describing an effective and broadly applicable decision framework.[1] Moreover, the committee has concluded that this literature and experience are broadly applicable to the examples considered and evaluated in this report.

The decision framework as applied to sustainability linkages must also include the major elements of relevant frameworks. These generally include the following elements:

- Agreement on the problem or issue and its scope
- Agreement on objectives and goals
- Agreement on "who's at the table"
- Engagement of all relevant stakeholders
- Capacity building to overcome asymmetries in stakeholder knowledge and resources

---

[1] In addition to the literature cited in Box 4-1, the World Bank has developed guidance for how to design a results framework, defined as "an explicit articulation (graphic display, matrix, or summary) of the different levels, or chains, of results expected from a particular intervention—project, program, or development strategy. The results specified typically comprise the longer-term objectives (often referred to as 'outcomes' or 'impact') and the intermediate outcomes and outputs that precede, and lead to, those desired longer-term objectives" (World Bank 2012).

BOX 4-1
Relevant Decision Framework Literature

Numerous National Academies reports include frameworks for decision-making on issues ranging from the environment to public health to transportation. Some selected reports include IOM 2010; NRC 1996, 2005, 2008a, b, 2009a, b, 2011a, b, 2012. In particular, NRC 2009b summarizes key issues related to decision support systems and distills six principles that are broadly related to the committee's framework that characterize these systems, including the benefits of following them. These include:

• **"Begin with users' needs.** Decision support activities should be driven by users' needs, not by scientific research priorities. These needs are not always known in advance, and they should be identified collaboratively and iteratively in ongoing two-way communication between knowledge producers and decision makers.

• **Give priority to processes over products.** To get the right products, start with the right process. Decision support is not merely about producing the right kinds of information products. Without attention to process, products are likely to be inferior—although excessive attention to process without delivery of useful products is also ineffective. To identify, produce, and provide the appropriate kind of decision support, interactions between decision support providers and users are essential.

• **Link information producers and users.** Decision support systems require networks and institutions that link information producers and users. The cultures and incentives of science and practice are different, for good reason, and those differences need to be respected if a productive and durable relationship is to be built. Some ways to accomplish this rely on networks and intermediaries, such as boundary organizations.

• **Build connections across disciplines and organizations.** Decision support services and products must account for the multidisciplinary character of the needed information, the many organizations that share decision arenas, and the wider decision context.

• **Seek institutional stability.** Decision support systems need stable support. This can be achieved through formal institutionalization, less-formal but long-lasting network building, new decision routines, and mandates, along with committed funding and personnel. Stable decision support systems are able to obtain greater visibility, stature, longevity, and effectiveness.

• **Design for learning.** Decision support systems should be structured for flexibility, adaptability, and learning from experience" (NRC, 2009b).

• Mutually negotiated and agreed upon decision rules (e.g., "how much agreement is sufficient to constitute approval") to ensure perceived legitimacy and accountability (may or may not require unanimity)

• Clarification of participant roles, responsibilities, and accountability

• Boundary processes/organizations at the intersection of scientists/technical experts and decision makers, managers, and stakeholders

- Maintenance of flexibility to adapt to new information and/or changing circumstances
- Understanding of structural barriers that could limit success and ways to address them.

A DECISION FRAMEWORK

Figure 4-1 presents a graphic representation of the decision framework recommended by the committee. The purpose of this framework is to lay out a structured but flexible process, from problem formulation through achievement of measureable outcomes, which engages agencies and stakeholders in goal-setting, planning, knowledge building, implementation, assessment, and decision adjustments. It is designed to be used when addressing place-based sustainability challenges as well as in policy formulation and rulemaking. The framework incorporates an iterative (or incremental) process that yields solutions to a wide range of issues that vary in scope, characteristics, and time. As an iterative process, the framework can also be viewed as a learning tool that begins with problem formulation and includes knowledge regarding key drivers and their relationship to key stakeholders, as well as access to scientific knowledge regarding the connections among components of the system. The framework is consistent with and extends the sustainability framework developed for the U.S. Environmental Protection Agency (EPA) in the "Green Book" (NAS, 2011a). As per the statement of task, the decision framework presented here will help "examine the consequences, trade-offs, synergies, and operational benefits of sustainability-oriented programs. The decision framework will include social, economic, and environmental dimensions of sustainability."

The framework is depicted in four phases: (1) preparation and planning; (2) design and implementation; (3) evaluation and adaptation; and (4) long-term outcomes. A description of each of the framework elements is given below. The framework is meant to apply to the creation of a sustainability program (an ongoing, interagency effort such as a crosscutting program to support sustainable development in cities) and projects (single interagency efforts focused on a specific task, such as a project to design sustainable water use and agricultural production in the Great Plains Ogallala Aquifer).

Phase 1: Preparation and Planning

This phase has three major steps that need to occur before the actual program or project is designed. This important phase and its associated steps are often overlooked or done in an incomplete or piecemeal fashion. The examples and other research done by this committee found that this phase and its elements were critical to the success of sustainability programs and, if not done well, contributed to the demise of programs. Because of the importance of this phase, a more detailed view is provided in Figure 4-2.

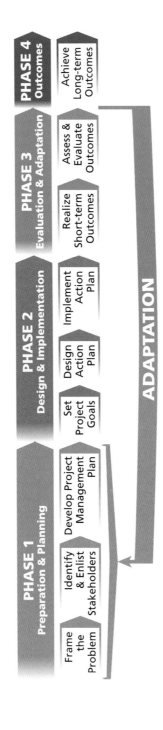

FIGURE 4-1 Conceptual Decision Framework. Four phases are shown, along with the relevant steps within each phase. The framework could be applied in creating either programs or projects related to sustainability.

75

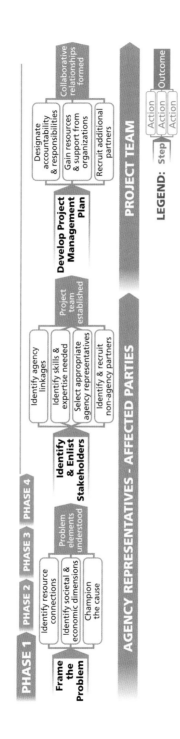

FIGURE 4-2 Phase 1 of the decision framework in expanded detail. Each step identified in Figure 4-1 of Phase 1 now includes specific actions and outputs/outcomes for that action (see key).

Phase 1: Preparation and Planning

This phase has three major steps that need to occur before the actual program or project is designed. This important phase and its associated steps are often overlooked or done in an incomplete or piecemeal fashion. The examples and other research done by this committee found that this phase and its elements were critical to the success of sustainability programs and, if not done well, contributed to the demise of programs. Because of the importance of this phase, a more detailed view is provided in Figure 4-2.

The steps that need to be taken in Phase 1, and their associated actions and outputs, include:

Frame the problem. A sustainability issue of sufficient complexity to warrant a multi-agency approach is first identified. Issues requiring a coordinated response are those of national significance due to their broad geographic extent, potential to impact long-term health and economic well-being, or crosscutting impact. Next, the issue must be framed so that the problem to be solved is clearly understood. This is analogous to problem formulation in human or ecological risk assessment. Effectively framing the problem requires a coordinated effort by an appropriate combination of federal, state, local, tribal, nongovernmental, and/or private-sector entities. An issue may be framed through a number of different avenues ranging from engaging key stakeholder partnerships to agency leadership and executive action. All dimensions of the problem must be identified, including the environmental resource connections, societal connections, and economic connections. These elements of the problem will inform the selection of agencies and nonagency organizations that should be involved in the program or project. It is important to note that agencies need not await structural overhauls in order to strengthen their capacity to address sustainability linkages. Agencies can begin by preparing a high-level systems map illustrating key linkages that can then be deployed widely across federal agencies for any sustainability-related program or project in order to incentivize policy coordination.

Some baseline analysis is typically required at this point to generally describe the magnitude of adverse impacts if the issue is not successfully addressed, and the magnitude of the benefits to be gained when it is. An initial estimate of the extent of the effort that might be reasonably expected to address the problem is also useful when framing it. These initial estimates will be refined as the decision process proceeds; thus, the process is iterative. An initial group of relevant parties—representatives of at least some of the relevant agencies, as well as some of the affected parties and those needed to implement potential solutions—are typically engaged at this point to assist with the framing. Some of these individuals often function as champions whose actions can engage relevant parties in the next step, as well as get buy-in from key agency administrators ("champion the cause").

Identify and enlist stakeholders. The next significant step is to identify the relevant agency linkages. Depending on the natural resources and social and economic aspects of the problem, it will be critical to engage all of the federal agencies affected by it. For example, a project to develop a sustainability plan for the Ogallala Aquifer would require participation by the U.S. Geological Survey (USGS), the Bureau of Reclamation (BOR), the Bureau of Land Management (BLM), the U.S. Fish and Wildlife Service (FWS), and the U.S. Department of Agriculture (USDA), as well as states, tribes, and others. This participation also illustrates the highly collaborative nature of the process, which continues throughout.

The issue framing conducted during the first step should provide knowledge as to which organizations and individuals need to be involved and the information needed to engage them. The interpersonal skills of the individuals engaged in the first, problem-framing step become critical in this phase, as they will often not have the positional authority to engage all of the relevant organizations and individuals. The initial group must collectively possess sufficient collaborative leadership skills to engage the relevant parties. It may also be necessary for them to identify and engage sponsors who have the influence to bring relevant parties to the table, along with necessary resources to support the efforts of the team. At this step of the process, the technical skills and professional expertise needed to design and implement the program or project are identified.

Identifying relevant nonagency stakeholders is part of this step as well. Nonagency stakeholders are frequently those who must use or implement the approach or solution developed to address the problem, as well as those impacted by the approach or solution. These stakeholders may be individuals or entities at the local, state, tribal, regional, or national scales. They may include nonfederal governmental agencies, nongovernmental organizations (NGOs), private-sector interests, or others who have significant interests in the outcome of decisions and actions. It is critical that all relevant players be involved; if a representative of a sector that is a key driver in the issue is missing, the likelihood of success is greatly diminished.

In the next step, the actual Project or Program Team ("Team") is identified. The Team, which may be deployed either to design a sustainability program or to address a specific sustainability problem at the project level, should include individual representatives from the relevant organizations ("stakeholders") identified during step 2. This group must have the necessary background, experience, and leadership skills to successfully design the project or program. Team members must be carefully selected by their member organizations; they must have the right commitment, expertise, and skill sets, and they must have appropriate authority from their organizations so that their participation leads to success. Sufficient expertise in the fields of environmental science, ecology, social science, economics, and public health should generally be included. Each Team member should be a collaborative leader, and each should add value to the Team. Members must be provided support and resources by their respective organizations. Attention should be paid to the informal and formal relationships that already exist across these organizations, as success can be strongly influenced by the trust that exists or is built among Team members.

If it has not been done previously, it is essential during this step to determine and specify the role the federal agencies will play relative to the other players. The agencies may be principal leaders, or facilitators, or deal-makers, or they may act as a backstop using their legal authorities, with regional, state, or other participants taking the leadership role. In several examples studied by the committee, federal agencies successfully provided (legal) cover for regional or local programs. Other successful examples highlighted federal agencies in a leadership role. Often it was the scale of the program (city vs. interstate) and willingness of the federal agencies to partner with and engage stakeholders effectively when they were in the position of leadership that contributed to success.

Develop project management plan. The importance of this step, in which the Team develops a management plan for the program or project, cannot be overstated. The plan should clearly delineate the roles, responsibilities, and accountability of each member organization or participant, as well as a business plan for the funding of the project design, implementation, and maintenance (thus assuring its longevity). Other partners may be identified at this point whose involvement will be necessary in order to meet the project goals and to balance any asymmetries in the capacity of the Team. This plan should be developed prior to any project design or implementation so as to avoid missing critical pieces and to avoid conflict among players as to who does what.

Phase 2: Design and Implementation

A more detailed version of Phase 2 is shown in Figure 4-3.

Set project goals. The Team establishes goals for the program or project—a step that should be taken with engagement of stakeholders and relevant members of the public. In addition, the short- and long-term outcomes and their associated measures are identified, and an evaluation process is developed. A project timeline for measuring and achieving goals is agreed upon. Goal and outcome settings may also inform the partnerships needed to achieve success. Evaluating baseline conditions before implementing a sustainability solution or approach is necessary so that a future evaluation can gauge the impact of the program.

Design action plan. Now the Team develops a comprehensive design of the approach, strategies, actions, etc. that are needed to address the sustainability issue and meet the goals established in the previous step. The necessary tools, knowledge, and information to accomplish the goals must be identified and pursued. The Team also needs to identify who will implement the plan, how the program will be maintained, and by whom. This plan must include "decision

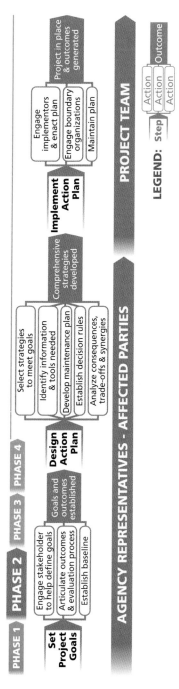

FIGURE 4-3 Phase 2 of the decision framework in expanded detail. Each step identified in Figure 4-1 of Phase 2 now includes specific actions and outputs/outcomes for that action (see key).

rules" (e.g., consensus or majority vote) for what constitutes acceptable actions or outcomes. The Team must build in the principles of adaptive management— that is, provide for flexibility in altering goals, design, and implementation as knowledge is gained in assessing the course of the program implementation and short-term outcomes (e.g., CRS, 2011).

It is critical to include a systematic and explicit process for projecting the outcomes of the program or project in order to anticipate the consequences, both intended and unintended, added benefits in terms of efficiencies and cost-savings, the short- and long-term trade-offs of implementing the plan (vs. doing nothing), and any synergies gained from the program or project. This can be done with a variety of tools, including scenario analysis (Schmitt Olabisi et al., 2010) and policy analysis (Bardach, 2012).

Implement action plan. The design phase is where the action plan is de-veloped for addressing the sustainability issue that was identified in the first step of Phase 1. This includes the "what" that needs to be done as well as the "how" and "by whom." It also includes a determination of the kinds of decision-making tools or models that might be needed for implementation. In this step, the action plan developed in the previous step is actually implemented, either by the Team that designed the program or project or by the implementers determined during the design step. A key action in this step is determining the kinds of boundary organizations or processes that are needed. (A boundary organization or process is one that bridges the scientific and technical people with the policy people and stakeholders either within or across entities, horizontally or vertically. Such or-ganizations often facilitate ongoing dialogue among experts and others (Guston et al., 2010).

Approaches to sustainability challenges generally take time and require maintenance to ensure their longevity, adoption, and success. The Team must develop and implement a maintenance plan that describes who is responsible for long-term maintenance, who pays for it, and who evaluates its effectiveness.

Phase 3: Evaluation and Adaptation

Realize short-term outcomes, assess outcomes, and adjust. This is where the "rubber meets the road" as results are achieved. Outcomes are assessed and evaluated relative to the baseline established in Phase 2. Short-term outcomes are on the scale of a year to a few years. Are the trends observed on track with goals? Significant learning typically occurs during this step as knowledge and actual experience are obtained, which allow modifications to framing the prob-lem, the approach, design, and methods. At this point the evaluation plan identi-fied above becomes critical, because it allows actual results to be compared to the original goals and for adjustments to be made. Additional stakeholders may also be identified and engaged at this point.

Phase 4: Long-Term Outcomes

Long-term outcomes are on the scale of several years or more, and should closely track the goals. While performance is assessed and adjustments are made during this phase, as in the previous one, a point is reached where a formal assessment is needed. Using the outcome measures developed under Phase 2, at this stage evaluations are conducted to see if short- and long-term outcomes are meeting goals. Ideally, the results of this evaluation should be able to be compared to the results of the baseline evaluation conducted in Phase 2. Based on this evaluation, necessary changes to the Team, Goals, Outcomes and Measures, Management Plans, Design, Implementation, or Maintenance are made.

When well executed, this framework process will enhance legitimacy, encourage systems thinking and the relevance of government actions, and most importantly, result in streamlined and more efficient governance. An additional benefit is that the experiences and lessons learned while applying this process are fed back to the participating organizations and individuals, improving both future efforts and government efficiency.

Finally, a decision framework for sustainability is unlikely to lead to consistently favorable actions unless several additional elements are also in place. An important factor is building sustainability into the fabric of an organization: its mission statement, its goals and objectives, and its organizational and management structure. A previous NRC report (2011a) that addressed sustainability at EPA discussed the importance of incorporating sustainability into an agency's culture and thinking. This committee (NRC, 2011a) found that integrating sustainability into the agency's work and culture will be most effective when based on clear principles, vision, strategic goals, and implementation processes. Also, the report recommended that the agency institute a focused program of change management to achieve the goal of incorporating sustainability into all agency thinking to optimize the social, environmental, and economic benefits of its decisions, and create a new culture among all EPA employees. Similarly, this committee found the incorporation of a culture of sustainability within the operations of the agency is essential. Also very important are structuring sustainability decision making on long time frames and assessing ways to maximize benefits in all sustainability solutions and approaches.

RECOMMENDATIONS

1. Federal agencies should adopt or adapt the committee's decision framework described above. Several key elements of the framework include the need to:
   a. Build sustainability into the fabric of an organization.
   b. Structure sustainability decision making on long time frames, incorporating adaptive management approaches.

    c. Assess co-benefits and trade-offs in all sustainability solutions and approaches, and communicate these along with the primary outcomes.

    d. Engage locals, states, and NGOs through an iterative processes to the extent possible, stressing inclusiveness, receptiveness, and good communications.

2. Agencies need not await structural overhauls in order to strengthen their capacity to address sustainability linkages. Agencies can begin by preparing a high-level systems map illustrating key linkages, which can then be deployed widely across federal agencies for any sustainability-related program or project in order to incentivize policy coordination.

REFERENCES

Bardach, E. 2011. A Practical Guide for Policy Analysis. 4th ed. Washington, DC: CQ Press.

EPA (U.S. Environmental Protection Agency). 2012. EPA Announces Framework to Help Local Governments Manage Stormwater Runoff and Wastewater. Online. Available at http://yosemite.epa.gov/opa/admpress.nsf/0/AB2035971BAB1AD48 5257A1B006F25B1. Accessed September 4, 2012.

Fiksel, J. 2006. A framework for sustainable materials management. *Journal of the Minerals Metals & Materials Society* 58(8):15-22.

Congressional Research Service. 2011. Adaptive Management for Ecosystem Restoration: Analysis and Issues for Congress.

Guston, D. H., W. Clark, T. Keating, D. Cash, S. Moser, C. Miller, and C. Powers. 2010. Report of the Workshop on Boundary Organizations in Environmental Policy and Science. New Brunswick, NJ: Bloustein School of Planning and Public Policy, Rutgers University.

IOM (Institute of Medicine). 2010. Bridging the Evidence Gap in Obesity Prevention: A Framework to Inform Decision Making. Washington, DC: National Academies Press.

Jabareen, Y. 2008. A new conceptual framework for sustainable development. *Environment, Development and Sustainability* 10:179-192.

Lockie, S., and S. Rockloff. 2005. Decision Frameworks: Assessment of the social aspects of decision frameworks and development of a conceptual model. Coastal CRC Discussion Paper. Norman Gardens, Australia: Central Queensland University.

NRC (National Research Council). 1996. Understanding Risk: Informing Decisions in a Democratic Society. Washington, DC: National Academies Press.

NRC. 2005. Decision Making for the Environment: Social and Behavioral Science Research Priorities. Washington, DC: National Academies Press.

NRC. 2008a. Research and Networks for Decision Support in the NOAA Sector Applications Research Program. Washington, DC: National Academies Press.

NRC. 2008b. Public Participation in Environmental Decision Making. Washington, DC: National Academies Press.

NRC. 2009a. Science and Decisions: Advancing Risk Assessment. Washington, DC: The National Academies Press.

NRC. 2009b. Informing Decisions in a Changing Climate. Washington, DC: National Academies Press.

NRC. 2009c. Sustainable Critical Infrastructure Systems: A Framework for Meeting 21st Century Imperatives. Washington, DC: National Academies Press.

NRC. 2011a. Sustainability and the U.S. EPA. Washington, DC: The National Academies Press.

NRC. 2011b. Improving Health in the United States: The Role of Health Impact Assessment. Washington, DC: National Academies Press.

NRC. 2012. Science for Environmental Protection: The Road Ahead. Washington, DC: National Academies Press.

Schmitt Olabisi, L. K., A. R. Kapuscinski, K. A. Johnson, P. B. Reich, B. Stenquist, and K. J. Draeger. 2010. Using scenario visioning and participatory system dynamics modeling to investigate the future: Lessons from Minnesota 2050. *Sustainability* 2(8):2686-2706.

Waheed, B., F. Khan, and B. Veitch. 2009. Linkage-based frameworks for sustainability assessment: Making a case for driving force-pressure-state-exposure-effect-action (DPSEEA) frameworks. *Sustainability* 2009(1):441-463.

World Bank. 2012. Designing A Results Framework for Achieving Results: A How-to Guide. Online. Available at http://siteresources.worldbank.org/EXTEVACAP DEV/Resources/designing_results_framework.pdf. Accessed February 28, 2013.

Chapter 5

A Path Forward: Priority Areas for Interagency Collaboration

Federal agencies face many challenges integrating decisions, both horizontally (across domains) and vertically (across federal, state, tribal, and local governments). Many of these challenges are neither new nor only recently identified.[1] At the same time, a number of interagency efforts, including some promising examples in various settings, have begun to successfully address sustainability linkages. Earlier chapters discuss some examples and, drawing from them, the committee developed principles and a framework for addressing interconnected issues and enhancing decision-making linkages among agencies, scientists, the private sector and the public. The committee recognizes that interconnections among issues are extensive, as are points of leverage to enhance interagency coordination. Thus, a challenge for agencies is determining where best to apply their efforts, both in terms of the relative importance of interconnected issues and the potential effectiveness of processes and practices that can strengthen interagency coordination.

This chapter identifies criteria for prioritizing sustainability issues that present significant connections among resource domains and across economic, social, and environmental dimensions. Using these criteria, this chapter highlights several priority issues that would benefit from the decision processes envisioned by this report's decision framework. This chapter also identifies some "bridging" areas that hold potential, in the near term, to strengthen interagency coordination and public-private collaboration as agencies and stakeholders grapple with the sorts of complex, interconnected issues described in this report.

---

[1] At least as far back as the 1950s, as expressed by the U.S. Commission on Intergovernmental Relations (Kestenbaum Commission), concerns about interagency coordination on interconnected issues have been raised (Kinkaid, 2011).

CRITERIA FOR SETTING PRIORITIES

The committee concludes that six criteria may be particularly relevant to identifying priority areas for addressing sustainability issues. These criteria build from the perspective that, while many challenges could benefit from some interagency coordination in research, goal-setting, and action, some challenges simply cannot be efficiently and effectively addressed without much more coordinated efforts, and it is these that should become priorities for the agencies.

The first criterion for priority selection is **national importance**. Every issue is important to someone, but some issues affect the nation; the inability to address these issues in a linked way among agencies and with the private sector can result in significant unintended consequences, duplicative effort, and high economic, environmental, and social costs.

A second criterion is the interdisciplinary nature of the issue. Issues that are **inherently interdisciplinary** would especially benefit from more integrated research and more coordinated action. For example, understanding the components and functions of ecosystems and the benefits they provide to human communities requires knowledge of biology, hydrology, geomorphology, air chemistry, human demographics, human consumption patterns, engineering, and so on.

A third criterion is the extent to which an issue involves **multiple interconnected resource domains**. For example, policies and practices to manage energy resources fundamentally affect and are affected by policies and practices regarding water, climate, air pollution, land use, biodiversity, public health, transportation, and other domains.

A fourth criterion is the degree to which agency research, policy, and action would **benefit from much greater multi-agency coordination**. At some level, every issue or resource domain is interconnected to others, and agencies and the private sector undertake actions that overlap, intersect, and sometimes compete. Yet some of these interconnected domains can nonetheless be reasonably managed without substantial interagency coordination. For others, sustainability fundamentally depends on much stronger interagency and public-private coordination to identify trade-offs, avoid unintended consequences and duplicated efforts, and ensure fairness in outcomes.

The fifth criterion is the potential for **leveraging private and civil society initiatives and resources**. Many interconnected sustainability challenges involve and impact the private sector and broader civil society. For example, considerable U.S. communications infrastructure is owned and operated by the private sector; energy production and distribution systems are largely privately owned; significant knowledge and capacity to respond to disasters resides among nonprofit organizations nationally and within local communities; and much scientific knowledge resides in universities and other nongovernmental organizations (NGOs). Effectively addressing sustainability linkages necessarily involves working with these potential partners to leverage their considerable knowledge, assets, and experiences, as well as to engage them in dialogue over goals and actions, resulting in significant multiplication of governmental actions to the benefit of the nation.

The sixth criterion is the prospect that applying the sustainability linkages decision framework and augmenting multi-agency, public-private sector coordination will have the potential to result in more effective, efficient outcomes *with positive return on investment* (either in the short term or, more likely, the long term). For issue clusters that involve decisions by multiple agencies, applying the framework can reduce duplication of effort and therefore potentially result in cost savings. It can also reduce unintended consequences in which actions taken to address one domain (for example, energy development) without considering other closely connected domains (for example, water and food supplies) can result in negative outcomes in those connected domains. Applying this criterion can help focus attention on those issue clusters in which many agencies have overlapping jurisdictions and in which the potential for unintended consequences is high.

PRIORITY DOMAINS AND ISSUE AREAS

Opportunities to better identify and address sustainability linkages are extensive.[2] The committee applied the selection criteria described above to highlight several significant issue clusters. All of these areas are nationally important, require interdisciplinary data and analysis, involve multiple interconnected resource domains, would benefit from greater coordination, have a potential to leverage nongovernmental knowledge and resources, and would result in positive returns on investment. Though they all demonstrate these characteristics, their areas of central focus vary.

1. Connections among Energy, Food, and Water

The availability of affordable supplies of energy, food, and water is vital to sustaining healthy populations and economic prosperity. Producing and using energy often involves consuming water and can also impact water quality, land use, air quality, and the agricultural sector. Producing ethanol to the 2012 target of 7.5 billion gallons per year was estimated by the U.S. Geological Survey (USGS) to require 30 billion gallons of water to process—the equivalent to the total water needs of Minneapolis. If a quarter of the corn crop used for ethanol requires irrigation, ethanol production will consume nearly a trillion gallons of water per year—equivalent to the combined water usage of all cities in Arizona,

---

[2]A 2012 NRC Symposium on Partnerships, Science, and Innovation for Sustainability Solutions included discussion of priority areas for the field (NRC, 2012). Additionally, a 1999 NRC report identifies eight priority areas needing greater attention and coordinated efforts to enhance sustainable outcomes that meet economic, social, and environmental goals.

Colorado, Idaho, and Nevada (DOI, 2006). In 2010, nearly 40 percent of U.S. corn was converted into ethanol, but the mandated amount of ethanol exceeds the supply, increasing the price of corn (Hanlon et al., 2013). Intensive corn production also has adverse environmental effects—chemical fertilizers that are heavily applied to corn crops cause run off, a major source of water pollution that affects drinking water. Likewise, some fossil fuel production, nuclear energy facilities, and renewable energy sources require water for production, processing, cooling, and other purposes.

A recent Government Accountability Office (GAO) Report noted that the energy sector is the fastest-growing consumer of water in the United States (GAO, 2012). Drawing upon a Congressional Research Service report, the GAO indicates that energy is "expected to account for 85 percent of the growth in domestic water consumption between 2005 and 2030." The reverse also applies—some water systems require large amounts of energy to transport water or treat it to necessary standards. Energy production and use are, thus, connected to water. But these connections are in turn affected by other factors. Population trends, urban infrastructure, agricultural production, and changes in economic activity all affect water demand. Complicating these connections is climate change and its effects on the availability and timing of water flows and on water temperatures and quality. Energy strategies therefore link to water management, infrastructure policy, and policies pertaining to climate change mitigation and adaptation. All of these factors—energy, water, and climate change—also affect food production and land use patterns. Better understanding of these connections, better coordination of federal agency actions, and enhancing public-private interactions to examine trade-offs and assess management strategies could improve economic, social, and environmental outcomes for the nation.

2. Diverse and Healthy Ecosystems

Ecosystems, their components, and functions provide "services" to human communities—for example, by supplying water, buffering against coastal storms, pollinating food-bearing plants, absorbing air pollution, and providing extractive minerals and other resources. While often not quantified, the economic value of these services represents a significant contribution to the economic health of the nation, and a significant economic burden would be added if these services were to disappear. The actions of many agencies affect these ecosystems, and many agencies and scientific disciplines contribute to better understanding these ecosystems and their functions. The inherently interconnected nature of ecosystem components requires an interdisciplinary approach to understanding and assessing the health of these ecosystems. Moreover, managing these ecosystems to sustain their benefits and long-term health often requires working at watershed or other larger scales; many different public and private land managers must work together to secure water quality along a river, for example, or to maintain dune systems that provide community protection against high-intensity storms. Sometimes capturing the economic value of these ecosystem services—sometimes referred to as "nature's capital"—requires that urban and nonurban areas and federal, state, local, and private-sector partners work together. For example, Denver's water utility is working with the U.S. Forest Service to invest in the removal of dead trees and overly dense vegetation in the area's watershed to reduce the prospects of a catastrophic wildland fire; such a fire could result in severe erosion and sediment that would damage the city's water reservoirs (U.S. Forest Service, 2011). Near Portland, Oregon, local water managers and the U.S. Department of Agriculture (USDA) are working to engage local farmers in planting trees along streams and rivers; the trees will provide shade, reducing water temperatures, benefiting fish habitat, and meeting water quality needs (Scarlett, 2010). These examples of joint management of connected domains are generating economic and social benefits, but these kinds of efforts remain relatively infrequent, suggesting that this is potentially an area of national significance that would benefit from continued and expanded focus.

3. Enhancing Resilience of Communities to Extreme Events

Sustainability of communities and regions is inherently tied to identifying and addressing vulnerabilities, promoting dynamic adaptation to change, and enhancing resilience in the face of disruptions (Fiksel, 2006). Disruptions may come from sudden catastrophic occurrences such as severe weather, earthquakes, or terrorist events, or from more progressive change such as that associated with a gradually warming climate.

A recent Department of Energy (DOE) technical report identified vulnerabilities associated with extreme events such as hurricanes or high-intensity rainfall events (DOE, 2012). The report particularly focuses on interdependencies and interconnections, noting that climate effects such as sea-level rise and storm surge can result in coastal flooding that in turn affects transportation, communications, water supplies, and energy services. Eyeing vulnerabilities to infrastructure, the report notes that "cross-sectoral issues related to infrastructures and urban systems have not received a great deal of attention; and, in fact, in some cases the existing knowledge base on cross-sectoral interactions and interdependencies...appears to be quite limited" (DOE, 2012, p. 1). While the DOE report significantly enhances understanding of these interdependencies, the nation's capacity to address them through coordinated multi-agency and public-private actions remains limited, as vividly demonstrated by Hurricane Sandy in late October, 2012. The National Response Framework discussed below clarifies roles and responsibilities across multiple agencies that need to coordinate actions in the wake of disasters and other emergencies. However, as its title suggests, the framework is focused on after-event responses rather than long-term infrastructure assessment and coordinated strategies to enhance resilience, reduce vulnerabilities, and meet infrastructure needs. There is a significant need to undertake such an assessment and develop more coordinated strategies for addressing vulnerabilities in infrastructure and promoting adaptation and resilience in communities.

Resilience has been defined as the capacity of a system to anticipate, prepare for, respond to, and recover from significant disruptions (Wilbanks and Kates, 2010); resilience allows a system to tolerate disturbance while retaining vital structure and function (Fiksel, 2003). Currently, our fundamental knowledge of what is required to enhance community resilience, whether in an urban, rural, or coastal environment, is inadequate. For example, what are the characteristics of communities that were more successful in quickly recovering from severe disturbances such as Hurricane Katrina? What roles do flexibility, dynamic adaptation, and infrastructure redundancy play? Many opportunities exist for collaboration among federal agencies in research, planning, strategy, and application to enhance resilience of communities to both sudden and ongoing stressors.

4. Human Health and Well-being

While Americans are in many ways healthier than ever, important health parameters continue to raise cause for concern. Some conditions have increased in prevalence over recent decades; examples include obesity (Center for Disease Control and Prevention (CDC), 2012a), asthma (Akinbami et al., 2012), diabetes (CDC, 2012b), autism spectrum disorders (Newschaffer et al., 2005), and some autoimmune diseases such as lupus (Uramoto et al., 1999). Mental health issues remain widespread: anxiety disorder affects 15 percent of people over the course of a lifetime and 10 percent in any year (Kessler et al., 2009); one in every 15 U.S. adults suffers a major depressive episode each year (Kessler et al., 2005); many more suffer from minor depression; and 9.4 percent of U.S. adults report "frequent mental distress" (Moriarty et al., 2009). Several trends—declining age at menarche (McDowell et al., 2007), rising prevalence of hypospadias (Paulozzi et al., 1997), and falling sperm counts (Swan et al., 2000)—may reflect exposure to endocrine disrupting chemicals (Wang et al., 2008; Nassar et al., 2010; Meeker et al., 2010), a worrisome possibility given the widespread presence of synthetic organic chemicals in tissue samples from the U.S. population (CDC, 2012c). Tens of thousands of Americans die in motor vehicle accidents each year, and hundreds die in severe weather events, which may be increasing in frequency. Importantly, many health impacts are unevenly distributed across the population. People of certain races, ethnic backgrounds, socioeconomic levels, ages, and disability status bear disproportionate risk in some circumstances, raising a range of equity concerns.

Sustainability efforts may affect each of these outcomes, and human health and well-being more generally, in complex, crosscutting ways. Agricultural practices affect the nutritional content and contaminant levels in food, as well as its availability and price. Land use and transportation decisions affect levels of physical activity, which in turn affect the risk of cardiovascular disease, many cancers, and other conditions. Transportation and energy decisions affect air quality, which affects the risk of cardiovascular and respiratory diseases. Urban design and preparedness efforts affect community resiliency, which in turn affects people's health and safety risks during and after disasters. Environmental policies affect health in numerous ways, including the probability of exposure to toxic chemicals, contaminated air and water, and hazardous waste. Housing—its availability, affordability, design, and quality—has a far-reaching impact on health and well-being, as do indoor environments in schools, workplaces, and health care facilities. Land conservation, from the scale of extensive wilderness areas to that of urban pocket parks, affects recreational opportunities, and biodiversity protection may facilitate future pharmaceutical development. These multifarious connections suggest that a linkages framework, engaging the many involved federal, state, and local government entities along with nongovernmental players—is essential to achieving sustainability policies that equitably and effectively promote human health and well-being.

IMPLEMENTATION BRIDGES

Agencies need not await policy or organizational restructuring in order to strengthen their capacity to address sustainability issues. Information provided by agencies, scholars, and research literature points to several ways to enhance agency capacity to operate with an interdisciplinary, cross-agency approach. In addition, examples presented to the committee illustrated ways to facilitate interagency coordination to address interconnected issues. This section highlights some of these "bridging" approaches, including roles and practices, legislative tools, and program planning.

Roles and Practices

Collaborative leadership: Collaboration and shared governance require more, not less, leadership (Emerson, 2012). Multiagency and public-private collaboration may generate conflict as different missions, values, purposes, and trade-offs become evident. Such conflict helps illuminate important issues and is, therefore, not a negative feature of collaboration. At the same time, managing these differences and conflicts requires effective leadership. One way to think about leadership competencies in collaborative settings is to consider three dimensions—attributes, skills, and behaviors—as described by Taylor and Morse (2011). For example, collaborative leadership requires systems thinking, facilitation of mutual learning, and building trusting relationships among partners.

Convening function: A first step in strengthening interagency and public-private sector coordination is to assemble diverse participants that have knowledge, resources, and interests related to a set of interconnected issues. Such convening can be initiated at all levels of government and by the nongovernmental and private sectors; however, federal agencies are often well-situated to serve in this convening role. The Desert Managers Group (DMG) in California emerged through federal leadership as the Bureau of Land Management (BLM) and other federal agencies recognized the crosscutting nature of the issues they faced. The group has no special implementation authority, but it has served as a forum to share information and discuss shared issues. In some circumstances, this sort of convening role can ultimately lead to more formal cross-agency, public-private sector decision structures, as it did in this instance.

Training and collaboration capacity: A common theme identified in all the examples reviewed by the committee was the importance of skills in collaboration, negotiation, and dialogue facilitation. These skills may be developed through mentoring programs for employees and cross-agency experiences. For example, a representative from the Federal Emergency Management Agency (FEMA), discussing the practice of emergency management, emphasized the need to reorient or retrain agency employees to think about the "problem statement," bringing a collaborative and community-centered approach rather than a hierarchical approach to emergency response (Kaufman, 2012).

Personnel competencies: As agencies have increasingly perceived the importance of interagency coordination and public-private collaboration, some agencies have pressed for more formal tools with which to assess relevant skills (Emerson, 2011). To this end, the Office of Personnel Management has added collaborative skills to its leadership competencies for senior executive service managers.[3]

Legislative Tools

As illustrated by many of the case studies assessed by the committee, federal agencies have access to several legislative and administrative tools that provide support for addressing interconnected issues and enhancing decision-making linkages among agencies. However, these tools are significantly underutilized. Below, the committee highlights two such tools—the recent regulatory and policy updates to the National Environmental Protection Act (NEPA) and statutory changes to the Government Performance and Results Act (GPRA)—that offer potential to strengthen federal capacity to address linked issues through interagency processes.

NEPA: Signed into law in 1970, NEPA provides a vision and general framework to link economic, social, and environmental aspects of federal agency decisions. It also anticipates the need for federal, state, and local coordination and sets forth provisions to facilitate such coordination. However, these provisions have generally been underutilized. A review of NEPA by the Council on Environmental Quality (CEQ) after 25 years of implementation concluded that

---

[3]See Office of Personnel Management's Leadership Competencies. Online. Available at: http://www.dtc.dla.mil/wfd/ldrshpdv/1.htm. Accessed October 24, 2012.

the Act's "most enduring legacy is as a framework for collaboration between federal agencies and those who will bear the environmental, social, and economic impacts of agency decisions" (CEQ, 1997 and 2007). A 2005 Memorandum of Agreement between Office of Management and Budget (OMB) and CEQ, which was reaffirmed in 2012 among 15 agencies, was intended to highlight agency collaboration under NEPA provisions by requiring a reporting of collaborative efforts. Several agencies have built upon NEPA's potential to enhance interagency coordination and collaboration. U.S. Department of the Interior's (DOI's) NEPA regulations, published in October 2008, provide directions for how to incorporate consensus-based management resulting from multi-participant collaboration into the NEPA process.[4] However, use of this and other tools has been limited.

GPRA: The importance of metrics in shaping actions and influencing outcomes is well-recognized in management literature (Melnyk et al., 2004). Though metrics are important, their utility depends on how well they actually capture and measure positive outcomes. GPRA, which requires agencies to develop performance measures, holds potential to motivate agencies to apply a sustainability framework and be accountable for coordinating with other agencies on shared and interconnected issues. Indeed, the amendments to GPRA in 2010 provide several changes that would support the inclusion of a focus on sustainability. For example, the revisions place "a heightened emphasis on priority-setting, cross-organizational collaboration to achieve shared goals, and the use and analysis of goals and measurement to improve outcomes" (Circular No. A-11, 2012, section 200-3). Notwithstanding this potential, drawing from presentations to the committee and a review of the literature, the committee notes that most agency GPRA measures have been narrowly focused and developed by individual offices for their particular programs. The committee supports the use of sustainability linkage metrics reflected in the GPRA Modernization Act guidance and urges the agencies to think more broadly in crafting their metrics.

Program Planning

Our case studies highlight several examples of resource connections and governance linkages in coastal, urban, and nonurban settings. Beyond these place-based examples, some agencies have developed broader programmatic initiatives that are designed to enhance governance linkages and collaborative partnerships.

One example is the Partnership for Sustainable Communities, an initiative started in 2009 by three federal agencies—the U.S. Environmental Protection

---

[4]The regulations state that, "in incorporating consensus-based management in the NEPA process, bureaus should consider any consensus-based alternative(s) put forth by those participating persons, organizations, or communities who may be interested in or affected by the proposed action" (DOI, 2008).

Agency (EPA), the Department of Housing and Urban Development (HUD), and the Department of Transportation (DOT). The purpose of the Partnership is to coordinate investments and align policies to "support communities that want to give Americans more housing choices, make transportation systems more efficient and reliable, reinforce existing investments, protect the environment, and support vibrant and healthy neighborhoods that attract businesses" (DOT, 2012). The Partnership develops programs and reviews grant applications to ensure that activities build on previous funding and meet multiple community goals. Since its inception, the Partnership has provided assistance to more than 700 communities all over the United States. In 2012, the Partnership identified several areas of focus, including continuing coordination to make government more efficient. One aspect of this coordination is offering joint training programs to help regional staff from the Partnership agencies develop knowledge and skills to support sustainable cities. The committee supports agency efforts to develop such programmatic initiatives.

A NATIONAL SUSTAINABILITY POLICY

Sustainability linkages are by their nature extraordinarily complex, involving multiple domains, multiple locations, and multiple time frames. The evidence-based studies described in this report highlight the national importance of sustainability efforts in urban, nonurban, and coastal environments. As discussed in Chapter 2, however, the federal government faces significant challenges in dealing with the inherent complexity of sustainability. The fragmentation of authorizations and appropriations for federal agencies, the lack of open access to necessary information and research results, and a government culture that reinforces silos have resulted in barriers to interagency coordination. Analysis of the examples described earlier in this report and consideration of additional reports and presentations received during the course of this study have led the committee to conclude that the success of complex, multiple-domain, interjurisdictional, multidisciplinary initiatives is significantly enhanced when addressed within the context of an overarching policy. Such a policy should clarify general goals and objectives, lay out governing principles, and provide for an operational/functional framework that explicitly delineates roles, authorities, and responsibilities. Examples and a discussion of this type of policy are given in the following text.

The lack of a guiding policy has limited the reach and effectiveness of collaboration in sustainability initiatives. In the absence of such a policy, agency participation in coordinated sustainability efforts has been uneven, capacity to develop unified or crosscutting budgets has been limited, and processes to develop shared goals on interconnected issues have been constrained. A national sustainability policy could significantly enhance the efficiency and effectiveness of complex initiatives involving multiple federal agencies, state, regional, and local governments, and nongovernmental stakeholders. The existence of such a policy would enable the development of institutional bridges, practices, or processes on which to build and maintain the necessary linkages among key responsible parties and stakeholders could be established under recognized policy authority, a statement of priorities, and established processes, leading to more successful and cost-effective sustainability efforts.

An Executive Order establishing a National Sustainability Policy and incorporating an implementation framework would substantially enhance the nation's capacity to address many of the governance challenges identified in Chapter 2. The objective of the National Sustainability Policy would be to address environmental, economic, and societal issues and support human well-being by: 1) encouraging and promoting coordination among agencies; 2) reducing siloed decision making and improving integration of research and operations across the government; 3) enhancing communication among agencies and between the federal government and stakeholders at national, state, and local levels; 4) reducing duplication of efforts and improving cost effectiveness; and 5) enhancing the use of existing laws such as NEPA by providing guidance on how to incorporate sustainability goals and linkages into federal decision making processes.

Several models exist for developing such a National Sustainability Policy, as discussed below:

The National Oceans Policy (NOP): Two recent reports that have galvanized presidential focus and action provide insights and recommendations on ocean policy. The first, released in 2003 by the PEW Center for the States, *Report to the Nation: Recommendations for a New Ocean Policy* (May 2003), called for national ocean policy legislation. The second, the U.S. Commission on Ocean Policy's 2004 report *An Ocean Blueprint for the 21st Century*, called for a coordinated and comprehensive national ocean policy. As a result of these reports, President Bush issued an Executive Order creating a Committee on Ocean Policy and calling for a U.S. Ocean Action Plan. In 2010 President Obama, building upon these reports and momentum, recognized stewardship of the oceans, coasts, and the Great Lakes as connected to national prosperity, environmental sustainability, and human well-being, and he signed an Executive Order developing a National Oceans Policy. The policy includes a set of overarching guiding principles for management decisions and actions toward achieving the vision of "an America whose stewardship ensures that the ocean, our coasts, and the Great Lakes are healthy and resilient, safe and productive, and understood and treasured so as to promote the well-being, prosperity, and security of present and future generations." According to the Administration, this policy will improve communication, coordination, and integration across all levels of government, and "agencies will streamline processes and reduce duplicative efforts, while better leveraging limited resources" (The White House, 2012).

The National Oceans Policy speaks to the need for connections similar to those required for sustainability in that it establishes a national framework to address a cross-governance challenge, and then engages stakeholders in regular meetings and other interactions designed to stimulate cooperative action. The committee views the NOP as a good model for addressing sustainability linkages.

National Incident Response Policy: FEMA is responsible for ensuring efficient and effective management of response to a wide range of incidents that invariably involve multiple domains and authorities. Since the founding of the agency, FEMA has learned through experience that the existence of a national policy and organizational framework greatly facilitates response to incidents. The two major policy structures aiding FEMA's work are the National Incident Management System and the National Response Framework. The components of these two policies are as follows:

- National Incident Management System:
    o A comprehensive, national approach to incident management
    o A template for incident management, regardless of size, location, or complexity
    o Application at all jurisdictional levels and across functional disciplines
- National Response Framework:
    o Guiding principles that enable response partners to prepare for and provide for a unified national response to all domestic disasters and emergencies
    o Application across all federal agencies and in coordination with state, local, and tribal agencies and the private and nongovernmental sectors.
    o Clarification of roles, responsibilities, and conditions for activating a unified response

Although these policy frameworks are not prescriptive, they present a common set of unifying principles and a structure for assessing and addressing complex, multiparty actions that promote efficient and effective outcomes. The policies guide efforts of all levels of government, the private and nonprofit sectors, and the public. The frameworks include guidance for planning, organization, and training needed to build and maintain domestic capabilities in support of the National Preparedness Goal. This is done, in part, through the development of a series of integrated national planning frameworks covering prevention, protection, mitigation, response, and recovery.

Desert Renewable Energy Conservation Plan (DRECP): The DRECP, discussed in Chapter 3 under the Mojave Desert example, was created in 2011 to help provide for effective protection and conservation of desert ecosystems while allowing for the appropriate development of renewable energy projects. The DRECP will provide long-term endangered species permit assurances to renewable energy developers and provide a process for conservation funding to implement the plan. It will also serve as the basis for one or more Habitat Conservation Plans under the Federal Endangered Species Act.

The DRECP was established by California law and a subsequent executive order from the governor (Senate Bill No. 2X [Joe Simitian, 2011-2012 1st Ex. Sess.], signed into law by Governor Brown on April 12, 2011) (DRECP, 2012). To oversee its implementation, a Renewable Energy Action Team (REAT) was formed, consisting of the California Energy Commission (CEC), California Department of Fish and Game (CDFG), the Bureau of Land Management (BLM), and the U.S. Fish and Wildlife Service (FWS). Memorandums of Understanding were signed by the participating agencies. Others joining the team include the California Public Utilities Commission, California Independent System Operator, National Park Service (NPS), EPA, and the Department of Defense (DOD) (California Department of Fish and Game et al., 2008).

Four major products and a schedule for their completion are being developed under the direction of the REAT:

1. Best Management Practices and Guidance Manual: Desert Renewable Energy Projects.
2. The Draft Conservation Strategy, which clearly identifies and maps areas for renewable energy project development and areas intended for long-term natural resource conservation as a foundation for the DRECP.
3. DRECP: a joint state and federal Natural Communities Conservation Plan and part of one or more Habitat Conservation Plans.
4. DRECP: draft and final joint state and federal Environmental Impact Report/Environmental Impact Statement.

Independent science advisors provided input into the Conservation Strategy and the DRECP. These advisors also completed the final report. Additional science input is expected as the process moves forward.

A stakeholder committee has been established to inform the state and federal REAT agencies on the development of the DRECP and to provide a forum for public participation and input. The stakeholders represent the interests of the counties in the desert region, renewable energy developers, environmental organizations, electric utilities, and Native American organizations. Specific working groups composed of DRECP stakeholder committee members have been established and meet regularly to address specific issues such as covered species, covered activities, resource mapping, and cultural resources.

Interjurisdictional approaches in other countries. There are abundant examples of sustainability strategies throughout the world, several of which could serve as instructive models in the development of a U.S. national policy. More than 100 countries have established national sustainable development strategies and have reported on them to the United Nations Commission on Sustainable Development.[5] In the committee's opinion, some of the best models can be seen

---

[5]The United Nations Sustainable Development Knowledge Platform. 2012. National Reports by Topic: National Sustainable Development Strategies (NSDS). Online. Availa-

in the United Kingdom (HM Government 2005; Department of Environment, Food, and Rural Affairs, 2011), Canada (Environment Canada, 2010), and the Netherlands (RMNO, 2007; 2008). All of these national policies and strategies present long-term goals for a sustainable nation and consider the environment and natural resources, economic health, and social well-being. They all are structured as broad frameworks that outline how long-term goals are to be achieved and do not create prescriptive processes. Rather, they provide clarity and a framework for how the range of stakeholders across government and outside of government can work together to achieve common sustainability goals. Most importantly, each of these policy and strategy instruments are living documents with clear provisions and processes for updating, refocusing, and evolving based on new knowledge and changing times.

RECOMMENDATIONS

This chapter has described criteria for prioritizing sustainability issues that present significant connections among resource domains and across economic, social, and environmental dimensions, and it highlights several priority issues that would benefit from the decision processes envisioned by this report's decision framework. "Bridging" areas have also been identified for strengthening agency coordination and public-private collaboration. It also discusses the need for a National Sustainability Policy. The following are key recommendations for action.

First, a National Sustainability Policy should be developed that will provide clear guidance to the executive agencies on addressing governance linkages on complex sustainability problems and inform national policy on sustainability. A process should be established for developing this policy, as well as a strategy for implementing it. All stakeholders, including the private sector and NGOs, should be provided an opportunity for contributing to this process. Once the policy is in place, agencies should develop specific plans to define how they expect to implement the policy. In implementing the National Sustainability Policy, consideration should be given to the creation of open and transparent oversight involving the public, state legislatures, Congress, and the President.

The committee suggests that an optimum National Sustainability Policy should be designed to accomplish the following:

1. Establish that the fundamental principle of sustainability is to promote the long-term sustainability of the nation's economy, environmental and natural resources, and social well-being.

---

ble at Available at http://sustainabledevelopment.un.org/index.php?menu=973. Accessed February 13, 2013.

2. Facilitate and empower sustainability initiatives across the federal government, including working with the many governmental and nongovernmental partners.

3. Set out broad general objectives, management principles, and a framework for addressing complex cross-jurisdictional sustainability challenges. However, it should not be prescriptive in approach, goals, participants, or structure.

4. Build sustainability and collaborative approaches that deal with sustainability connections into the fabric of governmental agencies.

The committee also believes it would be useful to clearly define the need for an initiative to enhance the ability of the federal government to address sustainability linkage issues; prepare an "initiative" communications kit to document the need for the initiative, its structure, goals, participants, etc.; and identify and communicate with key stakeholders and other audiences.

As discussed in Chapter 3, sustainability solutions need to be communicated in a way that clearly identifies both the costs and benefits of action and inaction. An effective communications strategy is important not only at the outset to engage major and important constituencies, but also throughout the process in keeping key stakeholders and the public generally aware of the progress being made and the work that still needs to be done. Research in this area will be important.

In addition, agencies should legitimize and reward the activities of individuals who engage in initiatives that "cross silos" in the interest of sustainability, both at the staff and leadership level. Among other things, agencies should develop personnel performance measures that emphasize collaboration and the design and implementation of interagency, integrated approaches to addressing sustainability issues. Agencies should nurture "change agents" both in the field and at regional and national offices, an effort that may include revisions to managers' performance plans, rewards, and training as well as better alignment of policy tools to support collaboration. Similarly, agencies should encourage and enable cross-agency management and funding of linked sustainability activities. In some cases, statutory authority to cross silos as well as to develop cross-agency funding on integrated cross-domain issues may be required.

Continuity in strategic plans that incorporate sustainability as a core value will require strong support from the highest levels of leadership. It is also necessary to maintain long-term initiatives on sustainability despite periodic temporal change in leaders (and changes in the beliefs and priorities of the leadership).

Agencies should also support long-term, interdisciplinary research underpinning sustainability. Among other things, the committee recommends funding robust research to provide the scientific basis for sustainability decision making. Sustainability challenges play out over long time scales; therefore, agencies should invest in long-term research projects on time scales of decades to provide the necessary fundamental scientific understanding of sustainability.

An example of such a long-term research program is the National Science Foundation's (NSF's) Long-Term Ecological Research (LTER) program. To successfully meet sustainability challenges, agencies will need to support additional interdisciplinary, cross-program research, such as NSF's Science, Engineering, and Education for Sustainability (SEES) Program. Although the impact of sustainability on human well-being is critically important, scientific information on this relationship is woefully inadequate and incomplete and needs to be strengthened at major health funding agencies, such as the National Institutes of Health. The committee also recommends a systematic analysis of network and governance models and adaptive decision making efforts to identify common issues and challenges.

Federal agencies that support scientific research should be incentivized to collaborate on sustained, cross-agency research. Sustainability should be supported by a broader spectrum of federal agencies, and additional federal partners should become engaged in science for sustainability. Federal agencies should collaborate in designing and implementing cross-agency research portfolios to better leverage funding.

It will also be critical to develop training for leadership and staff that includes both scientific and management aspects of sustainability issues and that addresses the system and agency linkages needed to achieve sustainability outcomes. Similar training should be incorporated into entry-level programs such as the Presidential Management Fellows program and into senior-level training such as the Senior Executive Service program.

The maintenance and enhancement of sustainability, a crosscutting issue vital to the United States over the long term, cannot afford to be constrained by fragmentation of authority, inadequate sharing of information, the structure of government, or other complexities. In this report, we suggest a number of approaches to minimize or surmount these challenges. It is important to the country to do so, and the committee hopes that its recommendations can be implemented with vigor and alacrity, for the linkages of sustainability in the federal government require it.

REFERENCES

Akinbami, L. J., J. E. Moorman, C. Bailey, H. S. Zahran, M. King, C. A. Johnson, and X. Liu. 2012. Trends in asthma prevalence, health care use, and mortality in the United States, 2001-2010. NCHS data brief 94. Hyattsville, MD: National Center for Health Statistics.

Baskin, L. S., T. Colborn, and K. Aimes. 2001. Hypospadias and endocrine disruption: is there a connection? *Environmental Health Perspectives* 109:1175-1183.

California Department of Fish and Game, California Energy Commission, California Bureau of Land Management, and the US Fish and Wildlife Service. 2008. Memorandum of Understanding between the California Department of Fish and Game, the California Energy Commission, the Bureau of Land Management, and the US Fish and Wildlife Service regarding the Establishment of the California Renewable

Energy Permit Team. Online. Available at http://www.blm.gov/pgdata/etc/media
lib/blm/ca/pdf/pa/energy.Par.76169.File.dat/RenewableEnergyMOU-CDFG-CEC-
BLM-USFWS-Nov08.pdf. Accessed August 31, 2012.

CDC (Centers for Disease Control and Prevention). 2012a. Obesity and Overweight.
Online. Available at http://www.cdc.gov/obesity/data/adult.html. Accessed December 31, 2012.

CDC. 2012b. Diabetes Date & Trends. Crude and Age-Adjusted Percentage of Civilian,
Noninstitutionalized Population with Diagnosed Diabetes, United States, 1980–
2010. Online. Available at http://www.cdc.gov/diabetes/statistics/prev/national/fig
age.htm. Accessed December 31, 2012.

CDC. 2012c. Fourth Report on Human Exposure to Environmental Chemicals, Updated
Tables. Atlanta: U.S. Department of Health and Human Services, Centers for Disease Control and Prevention. Online. Available at http://www.cdc.gov/exposure
report. Accessed December 31, 2012.

CEQ (Council on Environmental Quality). 1997. The National Environmental Policy Act:
A Study of Its Effectiveness After Twenty-five Years. Online. Available at http://
ceq.hss.doe.gov/nepa/nepa25fn.pdf. Accessed October 24, 2012.

CEQ. 2007. Collaboration in NEPA: A Handbook for NEPA Practitioners. Online. Available at http://ceq.hss.doe.gov/ntf/Collaboration_in_NEPA_Oct_2007.pdf. Accessed
February 15, 2013.

CEQ. 2012. National Ocean Council-National Ocean Policy Draft Implementation Plan.
Federal Register. Online. Available at https://www.federalregister.gov/articles/2
012/03/14/2012-6215/national-ocean-council-national-ocean-policy-draft-implem
entation-plan. Accessed August 9, 2012.

DEFRA (Department of Environment, Food, and Rural Affairs). 2011. Mainstreaming
Sustainable Development: The Government's Vision of what this Means in Practice. Online. Available at http://sd.defra.gov.uk/documents/mainstreaming-
sustainable-development.pdf. Accessed April 22, 2013.

DOE (U.S. Department of Energy). 2012. Climate Change and Infrastructure, Urban
Systems, and Vulnerability. *Technical Input to the U.S. National Climate Assessment*. Online. Available at http://www.esd.ornl.gov/eess/Infrastructure.pdf. Accessed October 24, 2012.

DOI (U.S. Department of the Interior). 2003. Environmental Statement Memorandum
No. ESM03-7. Online. Available at http://www.fedcenter.gov/_kd/Items/actions.
cfm?action=Show&item_id=2937&destination=ShowItem. Accessed October 24,
2012.

DOI. 2006. Water for America. Online. Available at http://www.doi.gov/budget/appro
priations/2009/upload/Water-for-America.pdf. Accessed October 23, 2012.

DOI. 2008. 43 CFR Part 46. Implementation of the National Environmental Policy Act
(NEPA) of 1969; Final Rule. *Federal Register* 73(200). October 15, 2008.

DRECP (Desert Renewable Energy Conservation Plan). 2012. Online. Available at
http://www.drecp.org/about/index.html. Accessed October 24, 2012.

Emerson, K. 2012. Presentation to the NRC Committee on Sustainability Linkages in the
Federal Government. Sixth Meeting. October 11, 2012.

Emerson, K., and L. Steven Smutko. 2011. Guide to Collaborative Governance. Policy
Consensus Initiative and University Network for Collaborative Governance.

Environment Canada. 2010. Planning for a Sustainable Future: A Federal Sustainable
Development Strategy for Canada. Gatineau, Quebec: Environment Canada.

Executive Office of the President and Council on Environmental Quality. 2011. Obama Administration's National Ocean Council Names State, Local and Tribal Representatives to Coordinating Body. Online. Available at http://www.whitehouse.gov/ad ministration/eop/ceq/Press_Releases/February_23_2011. Accessed August 9, 2012.

Fiksel, J. 2003. Designing resilient sustainable systems. *Environmental Science and Technology* 37(23):5330-5339.

Fiksel, J. 2006. Sustainability and resilience: toward a systems approach. *Sustainability: Science, Practice, and Policy* 2(2):14-21.

General Accounting Office. 2012. Energy-Water Nexus: Coordinated Federal Approach Needed to Better Manage Energy and Water Tradeoffs. *Report to the Ranking Member, Committee on Science, Space, and Technology.* U.S. House of Representatives.

Getha-Taylor, H., and R. Morse. 2011. Leadership development for local government executives: Balancing existing commitments and emerging needs. Presented at the 2011 Public Management Research Conference, June 4, 2011. Syracuse, NY: Syracuse University.

Hanlon, P., R. Madel, K. Olson-Sawyer, K. Rabin, and J. Rose. 2013. Food, Water and Energy: Know the Nexus. New York, NY: GRACE Communications Foundation. Online. Available at http://www.gracelinks.org/media/pdf/knowthenexus_final.pdf. Accessed February 25, 2013.

HM Government. 2005. One Future – Different Paths, The UK's Shared Framework for Sustainable Development.

Hoff, H. 2011. Understanding the Nexus: Background Paper for the Bonn2011 Conference on the Water, Energy and Food Security Nexus. Stockholm: Stockholm Environment Institute. Online. Available at www.water-energy-food.org/documents/ understanding_the_nexus.pdf. Accessed February 25, 2013.

Interagency Climate Change Adaptation Task Force. 2011. Draft National Action Plan: Priorities for Managing Freshwater Resources in a Changing Climate. Online. Available at http://www.whitehouse.gov/sites/default/files/microsites/ceq/napdraft 6_2_11_final.pdf. Accessed September 4, 2012.

Kessler, R. C., S. Aguilar-Gaxiola, J. Alonso, S. Chatterji, S. Lee, J. Ormel, T. B. Ustün, and P.S. Wang. 2009. The global burden of mental disorders: an update from the WHO World Mental Health (WMH) surveys. *Epidemiol Psychiatr Soc* 18(1):23-33.

Kessler, R. C., W. T. Chiu, O. Demler, K. R. Merikangas, and E. E. Walters. 2005. Prevalence, severity, and comorbidity of 12-month DSM-IV disorders in the National Comorbidity Survey Replication. *Archives of General Psychiatry* 62:617-627.

Kinkaid, J. 2011. U.S. Commission on Intergovernmental Relations: Artifact of a bygone era. *Public Administration Review* 71(2):181-189.

McDowell, M. A., D. J. Brody, and J. P. Hughes. 2007. Has age at menarche changed? Results from the National Health and Nutrition Examination Survey (NHANES) 1999-2004. *Journal of Adolescent Health* 40(3):227-231.

Meeker, J. D., S. Ehrlich, T. L. Toth, D. L. Wright, A. M. Calafat, A. T. Trisini, X. Ye, and R. Hauser. 2010. Semen quality and sperm DNA damage in relation to urinary bisphenol A among men from an infertility clinic. *Reproductive Toxicology* 30(4): 532-539.

Moriarty, D. G., M. M. Zack, J. B. Holt, D. P. Chapman, and M. A. Safran. 2009. Geographic patterns of frequent mental distress: U.S. adults, 1993–2001 and 2003–2006. *American Journal of Preventive Medicine* 46:497-505.

NSF (National Science Foundation). 2011. Long-Term Ecological Research Program: A Report of the 30 Year Review Committee. Online. Available at http://portal. nationalacademies.org/portal/server.pt/gateway/PTARGS_0_412321_4829_954_ 425760_43/collab/docman/download/337538/0/0/0/LTER-A%20Report%20of% 20the%2030%20Year%20Review%20Committee.pdf. Accessed October 4, 2012.

Nassar, N., P. Abeywardana, A. Barker, and C. Bower. 2010. Parental occupational exposure to potential endocrine disrupting chemicals and risk of hypospadias in infants. *Occupational and Environmental Medicine* 67:585-589.

Newschaffer, C. J., M. D. Falb, and J. G. Gurney. 2005. National Autism Prevalence Trends from United States Special Education Data. *Pediatrics* 115(3):e277-e282.

Natural Resources Conservation Service. Conservation Beyond Boundaries: NRCS Landscape Initiatives. Power point presentation, available from the NRCS, Office of the Chief, undated.

NRC (National Research Council). 1999. Our Common Journey. Washington, DC: National Academies Press.

NRC. 2012. Symposium on Partnerships, Science, and Innovation for Sustainability Solutions: A Meeting Summary. Online. Available at http://sites.nationalacademies. org/PGA/sustainability/index.htm. Accessed March 4, 2013.

OMB (Office of Management and Budget). 2012. Circular No. A–11. Preparation, Submission, and Execution of the Budget. Online. Available at http://www.whitehouse.gov/ sites/default/files/omb/assets/a11_current_year/a_11_2012.pdf. Accessed October 24, 2012.

OPM (Office of Personnel Management). Leadership Competencies. Online. Available at http://www.dtc.dla.mil/wfd/ldrshpdv/1.htm. Online. Accessed on October 24, 2012.

U.S. Department of Transportation. 2012. Partnerships for Sustainable Communities. Online. Available at http://www.sustainablecommunities.gov/aboutUs.html. Accessed October 24, 2012.

Paulozzi, L. J., J. D. Erickson, and R. J. Jackson. 1997. Hypospadias trends in two US surveillance systems. *Pediatrics* 100:831-834.

RMNO (The Netherlands Environmental Assessment Agency). 2007. A new Sustainable Development Strategy, June 2007.

RMNO. 2008. The Netherlands in a Sustainable World.

Scarlett, L. 2010. Green, Clean, and Dollar Smart: Ecosystem Restoration in Cities and Countryside. New York, NY: Environmental Defense Fund.

Scarlett, L. 2012. National Environmental Policy Act: Enhancing Collaboration and Partnerships. *Policy Commentary*. Washington, DC: Resources for the Future.

Swan, S. H., E. P. Elkin, and L. Fenster. 2000. The question of declining sperm density revisited: an analysis of 101 studies published 1934-1996. *Environmental Health Perspectives* 108:961-66.

United Nations Sustainable Development Knowledge Platform. 2012. National Reports by Topic: National Sustainable Development Strategies (NSDS). Online. Available at Available at http://sustainabledevelopment.un.org/index.php?menu=973. Accessed February 13, 2013.

Uramoto, K. M., C. J. J. Michet, J. Thumboo, J. Sunku, W. M. O'Fallon, and S. E. Gabriel. 1999. Trends in the incidence and mortality of systemic lupus erythematosus, 1950-1992. *Arthritis & Rheumatism* 42:46-50.

USDA (U.S. Department of Agriculture). 2012. Conservation Beyond Boundaries: NRCS Landscape Initiatives. Available at http://www.nrcs.usda.gov/initiatives/index.html. Online. Accessed October 24, 2012.

USDA Forest Service. 2011. Aurora Water and U.S. Forest Service Commit to Restore Hayman Fire Area. Online. Available at http://www.fs.usda.gov/detail/r2/news-events/?cid=STELPRDB5323468. Accessed October 23, 2012.

The White House. 2012. National Oceans Policy. Online. Available at http://www.whitehouse.gov/administration/eop/oceans/policy. Accessed August 29, 2012.

The White House. 2010. National Ocean Council. Online. Available at http://www.whitehouse.gov/administration/eop/oceans. Accessed August 9, 2012.

Wang, M. H., and L. S. Baskin. 2008. Endocrine disruptors, genital development, and hypospadias. *Journal of Andrology* 29(5):499-505.

Wilbanks, T., and R. Kates. 2010. Beyond adapting to climate change; embedding adaptation in responses to multiple threats and stressors. *Annals of the Association of American Geographers* 100(4):719-728.

Appendix A

Committee on Sustainability Linkages in the Federal Government

THOMAS GRAEDEL (NAE) (Chair) is the Clifton R. Musser Professor of Industrial Ecology, professor of chemical engineering, professor of geology and geophysics, and director of the Center for Industrial Ecology at Yale University. Previously, he was a distinguished member of the technical staff at AT&T Bell Laboratories. He has co-chaired the National Academies Roundtable on Science and Technology for Sustainability since 2008. He is the author or coauthor of 15 books and more than 350 technical papers in various scientific journals. Dr. Graedel received his B.S. in chemical engineering from Washington State University in 1960, his M.A. in physics from Kent State University in 1964, and his M.S. and Ph.D. in astronomy from the University of Michigan in 1967 and 1969, respectively. He was elected to the U.S. National Academy of Engineering in 2002 for "outstanding contributions to the theory and practice of industrial ecology."

ROBERT ANEX is a professor of biological systems engineering at the University of Wisconsin, Madison. He received a B.S. and M.S. in mechanical engineering and a Ph.D. in environmental engineering from the University of California, Davis. He previously held faculty positions in the Department of Agricultural and Biosystems Engineering at Iowa State University (2003-2010) and the Science and Public Policy Program and the School of Aerospace and Mechanical Engineering at the University of Oklahoma (1996-2003). Prior to his academic career, he was senior engineer and section head at Systems Control Technology, Inc., Palo Alto, California (1983-1991). He has served as a member of the EPA Board of Scientific Counselors, Science and Technology for Sustainability subcommittee. He served also on the National Research Council and National Academy of Sciences Committee on Science and Technology in Armenia. His research interests include assessing the environmental and economic performance of coupled human-natural systems, life cycle assessment, sustainability, and the impacts of biorenewable fuels and chemicals. He has served as an associate editor for the *International Journal of Life Cycle Assessment* and *Biotechnology for Biofuels*, and he currently serves as associate editor for the *Journal of Industrial Ecology*.

WILLIAM F. CARROLL is currently vice president of industry issues for Occidental Chemical Corporation and also adjunct professor of Chemistry at Indiana University, Bloomington. Dr. Carroll is a past president (2005) of the American Chemical Society and a current member of its board of directors. He is a fellow of the Royal Society of Chemistry, and chair or member of a number of committees for the National Research Council of the National Academy of Sciences. He is a member of advisory boards for DePauw University, Tulane University and is 2009 chair of the Council of Scientific Society Presidents. On behalf of OxyChem he has chaired numerous committees for industry associations, including the American Chemistry Council. He has served on expert groups commissioned by the United Nations Environment Programme, the U.S. Environmental Protection Agency and three states, and most recently the California Green Ribbon Science Panel. He has received the Henry Hill Award from the ACS Division of Professional Relations, the Michael Shea Award from the ACS Division of Chemical Technicians, an Indiana University Distinguished Alumni Service Award, and the Vinyl Institute's Roy T. Gottesman Leadership Award for lifetime achievement. He holds two patents and has over sixty publications in the fields of organic electrochemistry, polymer chemistry, combustion chemistry, incineration and plastics recycling. He holds a Ph.D. in organic chemistry from Indiana University, Bloomington, Indiana.

GLEN DAIGGER (NAE) is senior vice president and chief technology officer at CH2M HILL with responsibility for the technology function for the firm's water businesses (water resources, water supply and treatment, wastewater). He is also the first technical fellow for the firm, an honor which recognizes the leadership that he provides for CH2M HILL and for the profession in the development and implementation of new wastewater treatment technology. Dr. Daigger has more than 30 years of experience in wastewater treatment plant evaluation, troubleshooting, and process design. His areas of expertise include biological wastewater treatment and treatment process design, in particular biological nutrient removal (both nitrogen and phosphorus), combined trickling filter and activated sludge systems, the use of biological selectors to control activated sludge bulking, and oxygen transfer. Between 1994 and 1996 he served as professor and head of the Environmental Systems Engineering Department at Clemson University. Dr. Daigger is a member of the American Society of Civil Engineers, American Water Works Association, Association of Environmental Engineering, International Water Association, Water Environment, as well as numerous other professional societies. Dr. Daigger received his Ph.D. in environmental engineering, his master of science degree in environmental engineering, and a bachelor of science degree in civil engineering all from Purdue University.

PAULO FERRÃO is a professor at Instituto Superior Tecnico (IST) at the Technical University of Lisbon, where he is cofounder and current director of IN+, Center for Innovation, Technology and Policy Research. He is currently the national director of the MIT-Portugal Program, the major international partnership on science and technology in Portugal, in the field of engineering systems. He also coordinates the field of Sustainable Technologies and Environmental Systems of the Institute for Systems and Robotics (ISR). Dr. Ferrão developed his academic career at IST in the Department of Mechanical Engineering. He has been a professor since 1985, when he joined as a trainee assistant in the Systems Section of the Department; he became assistant professor in the Applied Thermodynamics Section in 1988, assistant professor in 1993, associate professor in 2001, and full professor in 2010. His teaching activity spans disciplines such as thermodynamics; energy systems analysis; environment, energy and development policies. He currently teaches the disciplines of industrial ecology and energy and environment and has the responsibility for coordinating the group of subjects on "Planning and Sustainable Development" in the scientific area of "Environment and Energy." Dr. Ferrão received his Ph.D. in mechanical engineering (1993) and his master in energy transfer and conversion (1998) from IST.

HOWARD FRUMKIN is dean of the University of Washington School of Public Health. From 2005 to 2010, he was at the U.S. Centers for Disease Control and Prevention, first as director of the National Center for Environmental Health and Agency for Toxic Substances and Disease Registry (NCEH/ATSDR), and later as special assistant to the director for climate change and health. Before joining CDC he was professor and chair of the Department of Environmental and Occupational Health at Emory University's Rollins School of Public Health and professor of medicine at Emory Medical School. Dr. Frumkin previously served on the board of directors of Physicians for Social Responsibility (PSR), where he cochaired the Environment Committee; as president of the Association of Occupational and Environmental Clinics (AOEC); as chair of the Science Board of the American Public Health Association (APHA), and on the National Toxicology Program Board of Scientific Counselors. As a member of EPA's Children's Health Protection Advisory Committee, he chaired the Smart Growth and Climate Change work groups. He is the author or coauthor of over 180 scientific journal articles and chapters. Dr. Frumkin received his A.B. from Brown University, his M.D. from the University of Pennsylvania, his M.P.H. and Dr.P.H. from Harvard, his internal medicine training at the Hospital of the University of Pennsylvania and Cambridge Hospital, and his occupational medicine training at Harvard. He is board-certified in both internal medicine and occupational medicine and is a fellow of the American College of Physicians, the American College of Occupational and Environmental Medicine, Collegium Ramazzini, and the Faculty of Occupational Medicine of the Royal College of Physicians of Ireland.

SALLY KATZEN served as administrator of the Office of Information and Regulatory Affairs in the Office of Management and Budget (OMB) ('93-'98), and then as deputy director of the National Economic Council in the White House ('98-'99), and deputy director for management at OMB ('2000-'01) in the Clinton administration. She is currently a visiting professor at the New York University School of Law, having previously taught at the University of Michigan Law School, George Mason University Law School, George Washington University School of Law, University of Pennsylvania Law School, Smith College, Johns Hopkins University, and the University of Michigan in Washington Program. She is also currently a senior advisor at the Podesta Group in Washington, DC. Before her government service, she was a partner in the Washington D.C. law firm of Wilmer, Cutler & Pickering, specializing in regulatory and legislative matters. She graduated magna cum laude from Smith College and magna cum laude from the University of Michigan Law School, where she was editor in chief of the Law Review. While in private practice, she served in various leadership roles in the American Bar Association, including chair of the Section on Administrative Law and Regulatory Practice and two terms as D.C. Delegate to the House of Delegates of the ABA, served as president of the Federal Communications Bar Association, and was president of the Women's Legal Defense Fund. Following graduation from law school, she clerked for Judge J. Skelly Wright of the United States Court of Appeals for the District of Columbia Circuit. She also served in the Carter administration for two years as the general counsel of the Council on Wage and Price Stability in the Executive Office of the President.

ANNA PALMISANO recently retired from the U.S. Department of Energy (DOE), where she served as associate director of science for biological and environmental research. At DOE, she was responsible for an annual budget of $600 million supporting basic research in bioenergy, systems biology and genomics, and climate and environment science. She also has served as the deputy administrator for competitive programs in U.S. Department of Agriculture, where she led the National Research Initiative. Previously, she was a program manager in the Office of Biological and Environmental Research, where she developed and managed research programs in bioremediation, carbon cycling, and genomics. Dr. Palmisano has also served as a program manager and acting division director for biomolecular and biosystems sciences and technology in the Office of Naval Research. She cochaired the U.S.-European Commission Working Group for Environmental Biotechnology from 1995 to 2010 and led the Interagency Microbe Project from 2004-2006. Dr. Palmisano received a B.S. degree in Microbiology from the University of Maryland and the M.S. and Ph.D. degrees in biology from the University of Southern California. She was an Allan Hancock Fellow at the University of Southern California and a National Research Council Fellow in planetary biology at NASA-Ames Research Center. She currently works as a science and technology consultant in biotechnology and bioenergy, agriculture and environment, and competitive grantsmanship.

STEPHEN POLASKY (NAS) is the Fesler-Lampert Professor of Ecological/Environmental Economics at University of Minnesota. He received a Ph.D. in economics from the University of Michigan in 1986. He previously held faculty positions in the Department of Agricultural and Resource Economics at Oregon State University (1993-1999) and the Department of Economics at Boston College (1986-1993). Dr. Polasky was the senior staff economist for environment and resources for the President's Council of Economic Advisers 1998-1999. He was elected into the National Academy of Sciences in 2010. He was elected as a fellow of the American Academy of Arts and Sciences in 2009 and a fellow of the American Association for the Advancement of Science in 2007. His research interests include ecosystem services, natural capital, biodiversity conservation, sustainability, integrating ecological and economic analysis, renewable energy, environmental regulation, and common property resources. He has served as coeditor and associate editor for the Journal of Environmental Economics and Management, as associate editor for International Journal of Business and Economics, and is currently serving as an associate editor for Conservation Letters, Ecology and Society and Ecology Letters, and on the Editorial Board of Proceedings of the National Academy of Sciences.

LYNN SCARLETT is currently codirector of the Center for Management of Ecological Wealth, Resources for the Future, in Washington, D.C., and an environmental analyst focusing on climate change adaptation, environmental risk management, green business and infrastructure, energy and water issues, landscape-scale conservation, and science and decision making. In 2009, she was a distinguished visiting lecturer on climate change at the University of California Bren School of Environmental Science and Management. From 2005 to 2009 she was deputy secretary at the Department of the Interior where she chaired the department's Climate Change Task Force. Previously, Dr. Scarlett served four years as the department's assistant secretary for policy, management, and budget. She is a former president of the Reason Foundation and director for 15 years of the Reason Public Policy Institute, where she focused on environmental, land use, and natural resources issues. She is a former president of Executive Women in Government and was chair of the federal Wildland Fire Leadership Council. She also serves on the boards of the American Hiking Society, the National Wildlife Refuge Association, the National Parks Conservation Association, the Consensus Building Institute, and RESOLVE (nonprofit environmental dispute resolution), and is a trustee emeritus of the Udall Foundation. She received her B.A. and M.A. in political science from the University of California, Santa Barbara, where she also completed her Ph.D. coursework and exams in political science.

ROBERT STEPHENS founded and has served as president of the Multi-State Working Group on Environmental Performance (MSWG), a national coalition of representatives from government, business, nongovernmental organizations, and academic institutions in the U.S. working on transformative policies relating to the environment and sustainable development. Via his continued involvement with the MSWG, Dr. Stephens serves as the secretariat to the Best Practice Network for Sustainable Development (BPN) for the United Nations Environment Program, Division of Technology, Industry, and Economics. Dr. Stephens retired in July 2004 from the California EPA after 30 years of service, most recently as assistant secretary for environmental management and sustainability. In this position, Dr. Stephens was responsible for the development and implementation of programs leading to environmental policy innovation and sustainability in California. Over his career, Dr. Stephens also served as deputy director of the Department of Toxic Substances Control for Science, Pollution Prevention, and Technology and Chief of the Hazardous Materials Laboratory for the state of California. Dr. Stephens is the primary and/or coauthor of some 60 articles and book chapters on topics ranging from basic environmental science and risk assessment to public policy related to the environment and sustainability. Dr. Stephens holds a Ph.D. in chemistry from the University of California and has held prior positions in industry and academia.

DEBORAH SWACKHAMER is professor and Charles M. Denny Jr., Chair in Science, Technology, and Public Policy in the Hubert H. Humphrey Institute of Public Affairs at the University of Minnesota, and codirector of the University's Water Resources Center. She also is professor in environmental health sciences in the School of Public Health. She received a B.A. in chemistry from Grinnell College and an M.S. and Ph.D. from the University of Wisconsin-Madison in water chemistry and limnology & oceanography, respectively. After two years of postdoctoral research in chemistry and public & environmental affairs at Indiana University, she joined the Minnesota faculty in 1987. Dr. Swackhamer currently serves as chair of the chartered Science Advisory Board of the U.S. Environmental Protection Agency, and on the Science Advisory Board of the International Joint Commission of the U.S. and Canada. She currently serves on the National Research Council, National Academy of Sciences committee reviewing the USGS National Assessment of Water Quality Program. She is appointed by Governor Pawlenty to serve on the Minnesota Clean Water Council. She is a fellow in the Royal Society of Chemistry in the UK.

LAUREN ZEISE is chief of the Reproductive and Cancer Hazard Assessment Branch of the California Environmental Protection Agency. She oversees or is otherwise involved in a variety of California's risk assessment activities, including cancer and reproductive toxicant assessments; development of frameworks and methodologies for assessing cumulative impact, nanotechnology, green chemistry and safer alternatives, and susceptible populations; the California Environmental Contaminant Biomonitoring Program; and health risk characterizations for environmental media, food, fuels, and consumer products. Dr. Zeise's research focuses on human interindividual variability, dose response, uncertainty, and risk. She was the 2008 recipient of the Society of Risk Analysis's Outstanding Practitioners Award and is a national associate of the NRC. She has served on various advisory boards and committees of EPA, Office of Technology Assessment, the World Health Organization, and the National Institute of Environmental Health Sciences. She has also served on numerous NRC and Institute of Medicine committees and boards, including the committees that produced *Toxicity Testing in the 21st Century: A Vision and Strategy; Science and Decisions: Advancing Risk Assessment;* and *Understanding Risk: Informing Decisions in a Democratic Society.* Dr. Zeise received her Ph.D. from Harvard University.

Appendix B

Statement of Task

An ad hoc committee under the Science and Technology for Sustainability Program will identify the linkages among areas such as energy, water, health, agricultural production, and biodiversity that are critical to promoting and encouraging long term sustainability within the federal policy framework, recognizing that progress towards sustainability necessarily involves other levels of government, the private sector, and civil society. The premise is that achieving sustainability (defined in the 2009 Executive Order 13514) is a systems challenge that cannot be realized by separately optimizing pieces of the system. The study will build upon existing and emerging expertise throughout the scientific and technological communities. It will describe the nexus where domains intersect but in which existing institutions and disciplines often do not intersect.

The committee will convene a series of fact finding meetings, commission expert-authored case studies, and review the pertinent literature. Based on the information gathered from these sources, the committee will produce a report with consensus findings that provides an analytical framework for decision making. This framework can be used by U.S. policy makers and regulators to assess the consequences and tradeoff/synergies of policy issues involving a systems approach to long term sustainability and decisions on sustainability-oriented programs. The framework will include social, economic and environmental dimensions of sustainability, highlighting certain dimensions that are sometimes left unaccounted for in cross media analyses.

The report will also:

- Recommend priority areas for interagency cooperation on specific sustainability challenges;
- Identify impediments to interdisciplinary, cross-media federal programs; and
- Highlight scientific research gaps as they relate to these interdisciplinary, cross-media approaches to sustainability.

Appendix C

Committee Meeting Agendas
Open Sessions

FIRST COMMITTEE MEETING
September 20, 2011
The Dupont Circle Hotel
1500 New Hampshire Avenue, Washington, DC 20036

10:15 am   *Welcome and Introductions*
               Thomas Graedel, Clifton R. Musser Professor of Industrial
               Ecology, Yale University; Chair of the National Research Council
               Committee on Sustainability Linkages in the Federal Government

10:30 am   *Charge from Sponsors*
               • Paul Anastas, Assistant Administrator, Office of Research
               and Development, U.S. Environmental Protection Agency
               • Kai Lee, Program Officer, Conservation and Science Program,
               The David and Lucile Packard Foundation (teleconference)
               • Thomas Grumbly, Vice President, Civil Government
               Programs, Lockheed Martin
               • Carl Shapiro, Senior Economist, U.S. Geological Survey
               • Andrew Szilagyi, Director, Office of Deactivation and
               Decommissioning, U.S. Department of Energy
               • James Leatherwood, Director, Environmental Management
               Division, National Aeronautics and Space Administration
               • Ann Bartuska, Deputy Under Secretary for Research,
               Education, and Economics, U.S. Department of Agriculture
               • Margaret Cavanaugh, Deputy Assistant Director for
               Geosciences, National Science Foundation

- Larry Robinson, Assistant Secretary of Commerce for Oceans and Atmosphere, National Oceanic and Atmospheric Administration
- Ellen Williams, Chief Scientist, BP
- Marilu Hastings, Program Director, The Cynthia and George Mitchell Foundation (teleconference)

12:20 pm	*General Discussion and Q&A on Sponsor Perspectives*
12:40 pm	LUNCH
1:30 pm	*State and Local Government Perspective on Sustainability Linkages* David Paylor, Director, Virginia Department of Environmental Quality Howard Neukrug, Commissioner, City of Philadelphia
2:10 pm	*Industry Perspective on Sustainability Linkages* Hank Habicht, Managing Partner, SAIL Capital Partners Edwin Pinero, Executive Vice President, Veolia Water North America (teleconference)
2:50 pm	BREAK
3:05 pm	*Nongovernmental Perspective on Sustainability Linkages* Rebecca Shaw, Associate Vice President, Environmental Defense Fund Marty Spitzer, World Wildlife Fund
3:45 pm	*National Perspective on Sustainability Linkages* Sally Ericsson, Director, Natural Resource Programs, Office of Management and Budget Bruce Rodan, Senior Policy Analyst, Office of Science and Technology Policy
4:25 pm	*Department of Defense Perspective on Sustainability Linkages* Richard G. Kidd IV, Deputy Assistant Secretary of the Army, U.S. Department of Defense
4:40 pm	**Open Microphone Session: Brief Comments from Interested Parties** *Comments will be limited to 5 minutes, if you would like to address the committee please send an email to sustainability@nas.edu*
5:15 pm	Open Session Adjourns

SECOND COMMITTEE MEETING
February 7-8, 2012
Burke Museum of Natural History and Culture
Located near the University of Washington Campus
17th Ave and 45th Street, NE

Tuesday, February 7, 2012

9:00 am	*Welcome and Introductions* Marina Moses, Science and Technology for Sustainability Program, National Academies Julie Stein, Executive Director, Burke Museum of Natural History and Culture
9:10 am	*Introduction to Linkages of Sustainability* Thomas Graedel, Committee Chair, Yale University
9:30 am	*Overview of Sustainability Linkages in Coastal Systems* Josh Baldi, Special Assistant to the Director, Washington State Department of Ecology

Opportunities and Challenges to Puget Sound as a Coastal System

Moderator: Robert Stephens, President, Multi-State Working Group on
Environmental Performance

10:00 am	Alan Durning, Executive Director and Founder, Sightline Institute
10:15 am	Mike Grady, Transportation Branch Chief and Federal Green Challenge, National Oceanic and Atmospheric Administration-NWR representative
10:30 am	BREAK
10: 45 am	Billy Frank, Jr., Chairman, Northwest Indian Fisheries Commission
11:00 am	Michael Rylko, Senior Technical Coordinator, Puget Sound National Estuary Program, EPA Region 10
11:15 am	Panel Discussion
12:00 pm	LUNCH BREAK

Energy in the Pacific Northwest

Moderator: Lisa Graumlich, Dean, College of the Environment, University of Washington

1:15 pm	Chuck Clarke, Chief Executive Officer, Cascade Water Alliance
1:30 pm	Angus Duncan, Founder and President, Bonneville Environmental Foundation
1:45 pm	K.C. Golden, Policy Director, Climate Solutions
2:00 pm	Phil Rockefeller, (Senator, D-Bainbridge Island), Northwest Power and Conservation Council
2:15 pm	Panel Discussion
3:00 pm	BREAK

The Built Environment, Land Use, and Public Health around Puget Sound

Moderator: Howard Frumkin, Dean, School of Public Health and Professor, University of Washington

3:15 pm	Gene Duvernoy, President, Cascade Land Conservancy
3:30 pm	Gerry O'Keefe, Executive Director, Puget Sound Partnership
3:45 pm	David Fleming, Director and Health Officer, Seattle and King County
4:00 pm	Panel Discussion
4:45 pm	**Open Microphone Session: Brief Comments from Interested Parties** *Comments will be limited to 5 minutes, if you would like to address the committee please send an email to sustainability@nas.edu*
5:00 pm	*Meeting Wrap-up* Thomas Graedel, Committee Chair, Yale University
5:15 pm	Meeting Adjourns

Wednesday, February 8, 2012

9:00 am   *Welcome and Introduction to Linkages of Sustainability*
                      Thomas Graedel, Committee Chair, Yale University

Opportunities and Challenges to the Great Lakes as a Coastal System

Moderator: Rich Moy, Commissioner, US Section of the International
Joint Commission (IJC)

9:15 am   Deborah L. Swackhamer, Professor, School of Public Health and
                      Co-Director, Water Resources Center, University of Minnesota

9:30 am   James P. Bruce, Co-Chair, Public Interest Advisory Group

9:45 am   Hugh MacIsaac, Professor, Department of Fisheries and Oceans
                      Invasive Species Research Chair, Director, Canadian Aquatic
                      Invasive Species Network (CAISN)

10:00 am   Roy Stein, Emeritus Professor, Aquatic Ecology Laboratory and
                      Department of Evolution, Ecology and Organismal Biology, The
                      Ohio State University

10:15 am   Panel Discussion

11:00 am   BREAK and Closed Session for Committee Discussion

Land Use, Built Environment, and the Great Lakes Regional Economy

Moderator: Victoria Pebbles, Program Director, Great Lakes Commission

1:00 pm   Jennifer Read, Acting Director, Michigan Seat Grant and
                      Executive Director, Great Lakes Observing System (via
                      teleconference)

1:15 pm   Jim LaGro, Department of Regional and Urban Planning,
                      University of Wisconsin

1:45 pm   Nancy Frank, Associate Professor, School of Architecture and
                      Urban Planning, University of Wisconsin – Milwaukee

2:00 pm   Panel Discussion

2:45 pm   BREAK

Management and Governance in the Great Lakes Region

Moderator: Dave Naftzger, Executive Director, Council of Great Lakes Governors

3:00 pm   Cameron Davis, Senior Advisor to the Administrator on Great
               Lakes, Environmental Protection Agency (via teleconference)

3:15 pm   Dave Naftzger, Executive Director, Council of Great
               Lakes Governors

3:30 pm   Victoria Pebbles, Program Director, Great Lakes Commission

3:45 pm   Panel Discussion

4:00 pm   **Open Microphone Session: Brief Comments from
               Interested Parties**
               *Comments will be limited to 5 minutes, if you would
               like to address the committee please send an email to
               sustainability@nas.edu*

4:45 pm   Meeting Wrap-up
               Thomas Graedel, Committee Chair, Yale University

5:00 pm   Open Session Adjourns

THIRD COMMITTEE MEETING
April 11-12, 2012
Courtyard Omaha Downtown/Old Market Area
101 South 10th Street Omaha, NE 68102

Wednesday, April 11, 2012

8:45 am   *Welcome and Introductions*
               Thomas Graedel, Committee Chair, Yale University
               Marina Moses, Science and Technology for Sustainability
               Program, National Academies

9:00 am   *Overview of Sustainability Linkages in Nonurban Systems*
               Lynn Scarlett, Resources for the Future

Introduction to Sustainability Linkages in the Mojave Region

9:30 am   *Introduction to the Mojave Region*
               Buford Crites, Former Mayor, Palm Desert, CA and
               Professor, College of the Desert

9:45 am   *Introduction to Sustainability Linkages in the Mojave Desert*
                 Laura Crane, Director, Renewable Energy Initiative, The
                 Nature Conservancy

10:00 am   Q&A

10:20 am   BREAK

Interagency Coordination – Desert Managers Group

Moderator: Robert Anex, Professor of Biological Systems, University of
Wisconsin-Madison

10:30 am   *Introduction to the Desert Managers Group*
                 Henri Bisson, Senior Advisor for Renewable Energy
                 Development, Marstel Day, LLC and former Deputy Director
                 for Operations, U.S. Department of the Interior

10:55 am   *Management across agency boundaries—obstacles and*
                 opportunities: USMC Perspective
                 Pat Christman, Director, Western Region Environmental Office,
                 U.S. Marine Corps

11:20 am   *Management across agency boundaries – obstacles and*
                 opportunities: DOI Perspective
                 Russell Scofield, Coordinator, Desert Managers Group,
                 U.S. Department of the Interior

11:55 am   *Panel response to questions from the committee*
                 Dennis Schramm, former Superintendent, Mojave National
                 Preserve, National Park Service (via teleconference)

12:25 pm   LUNCH

Interagency Coordination – Desert Renewable Energy Conservation
Plan (DRECP)

Moderator: Robert Anex, Professor of Biological Systems, University of
Wisconsin-Madison

1:25 pm   *Overview of the Desert Renewable Energy Conservation*
                 Plan (DRECP)
                 Karen Scarborough, Landscape Architect, Former
                 Undersecretary of the California Natural Resources Agency

1:50 pm   *Management across agency boundaries – obstacles
and opportunities*
Wayne Spencer, Lead Scientist and Science Facilitator to
DRECP, Conservation Biology Institute

2:15 pm   *Panel response to question from the committee*

2:45 pm   BREAK

3:00 pm   **Open Microphone Session: Brief Comments from
Interested Parties**
*Comments will be limited to 5 minutes, if you would
like to address the committee please send an email to
sustainability@nas.edu*

3:15 pm   *Meeting Wrap-up*
Thomas Graedel, Committee Chair, Yale University

3:30 pm   Open Session Adjourns

Thursday, April 12, 2012

8:45 am   *Welcome and Introductions*
Thomas Graedel, Committee Chair, Yale University
Marina Moses, Science and Technology for Sustainability
Program, National Academies

Introduction to Sustainability Linkages in the Platte River Basin

9:00 am   *Introduction to the Platte River Basin—Brief History*
David Freeman, Professor Emeritus, Colorado State University

9:35 am   *Sustainability Linkages, Complexities, and Uncertainties
in the Platte River Basin*
Chad Smith, Headwaters Corporation, Director of
Natural Resources

10:10 am   BREAK

Resource Linkages and Competing Uses: Integrating Science and Action

Moderator: Lynn Scarlett, Resources for the Future

10:30 am   *Adaptive Management as an Integrating Process*
Chad Smith, Director of Natural Resources,
Headwaters Corporation

10:55 am   *Integrating Science and Action: Fish & Wildlife*
                            Service Perspective
                            Mike Thabault, Assistant Regional Director, Ecological
                            Services, Mountain-Prairie Region

11:20 am   *Panel response to questions from the committee*

11:50 am   LUNCH

**Interagency Coordination—Platte River Recovery Implementation
Program and Governance Committee**

Moderator: Lynn Scarlett, Senior Visiting Scholar, Resources for the Future

12:45 pm   *Introduction: Platte River Recovery Implementation Program*
                            and Governance Committee
                            Jerry Kenny, President and CEO, Headwaters Corporation;
                            Executive Director, Platte River Recovery Implementation
                            Program

1:10 pm   *Management across agency boundaries—obstacles and*
                            opportunities: Bureau of Reclamation Perspective
                            Gary Campbell, Deputy Regional Director, Great Plains
                            Regional Office, Bureau of Reclamation

1:35 pm   *Management across boundaries – obstacles and opportunities:*
                            Water User Perspective
                            Don Kraus, Central Nebraska Public Power and Irrigation District

2:00 pm   *Lessons Learned—A Decade of Evolving Governance*
                            David Freeman, Professor Emeritus, Colorado State University

2:25 pm   *Panel responses to questions from the committee*

2:55 pm   BREAK

3:05 pm   **Open Microphone Session: Brief Comments from
                            Interested Parties**
                            Comments will be limited to 5 minutes, if you would
                            like to address the committee please send an email to
                            sustainability@nas.edu

3:15 pm   *Meeting Wrap-up*
                            Thomas Graedel, Committee Chair, Yale University

3:30 pm   Open Session Adjourns

FOURTH COMMITTEE MEETING
June 11-12, 2012
Arizona State University (ASU) Wrigley Hall
800 S. Cady Mall Tempe, AZ 85287

June 11, 2012

8:30 am   *Welcome and Introductions; Discussion of Matrix*
               *and Framing Questions**
               Thomas Graedel, Committee Chair, Yale University

"Plenary" Session: Urban Sustainability and Linkages

9:00 am   *Urban Sustainability: Overview of Key Issues Including Energy*
               Paulo Ferrao, Professor, Instituto Superior Tecnico,
               Technical University of Lisbon

9:30 am   *Forging Interagency Linkages on Sustainability*
               Xavier Briggs, Associate Professor of Sociology and
               Urban Planning, Massachusetts Institute of Technology

Urban Sustainability and Linkages: Phoenix

10:00 am   *Overview of Sustainability Challenges in Phoenix*
               Dan Childers, Professor, ASU School of Sustainability and
               Director, Central Arizona-Phoenix LTER

10:30 am   *Conservation of Threatened Species and Habitats*
               Kimberly McCue, Program Director, Conservation of
               Threatened Species and Habitats, Desert Botanical Garden

11:00 am   *Urban Heat Islands and Public Health*
               Diana Petitti, Professor, ASU's Department of Biomedical
               Informatics

11:30 am   *Panel Discussion*
               Moderator: Anna Palmisano, U.S. Department of Energy (retired)

12:00 pm   LUNCH

1:00 pm   *Water Resources and Sustainability in Phoenix*
               Ray Quay, ASU Decision Center for a Desert City and
               Former Assistant Director, Water Services, City of Phoenix

1:30 pm   *Design of the Urban Landscape*
               Billie Turner, Gilbert F. White Professor of Environment and
               Society, ASU's School of Geographical Sciences

2:00 pm	*Overview of Current Sustainability Efforts in Phoenix* Colin Tetreault, Senior Policy Advisor of Sustainability to Mayor Stanton
2:30 pm	*Panel Discussion* Moderator: Paulo Ferrao, Professor, Technical University of Lisbon
3:00 pm	*Meeting Wrap-up* Thomas Graedel, Committee Chair, Yale University
3:15 pm	Open Session Adjourns

June 12, 2012

7:00 am	Meet in lobby of hotel for field trip to tour Phoenix (led by Dan Childers and Stevan Earl, ASU)

ASU Wrigley Hall

10:15 am	*Welcome and Introductions; Discussion of Matrix and* *Framing Questions** Thomas Graedel, Committee Chair, Yale University

Urban Sustainability and Linkages: Philadelphia

10:30 am	*Overview of Sustainability Challenges in Philadelphia* Mark Alan Hughes, Distinguished Senior Fellow, University of Pennsylvania
11:00 am	*Public Health Issues and Sustainability in Philadelphia* Charles Branas, Associate Professor of Epidemiology, Director of the Cartographic Modeling Laboratory, University of Pennsylvania
11:30 am	*Panel Discussion* Moderator: Glen Daigger, Senior Vice President and Chief Technology Officer, CH2M HILL
12:00 pm	LUNCH
1:00 pm	*Transportation Issues and Sustainability in Philadelphia* Erik Johanson, Strategy and Sustainability Planner, Southeastern Pennsylvania Transportation Authority

1:30 pm	*Water Resources and Sustainability in Philadelphia* Christopher Crockett, Director of Planning & Research, Philadelphia Water Department (via teleconference)
2:00 pm	*Panel Discussion* Moderator: Deborah Swackhamer, Professor, University of Minnesota
2:30 pm	*Meeting Wrap-up* Thomas Graedel, Committee Chair, Yale University
3:00 pm	Open Session Adjourns

*Framing Questions
1. What are the barriers to linkages, integration and synergy among agencies supporting urban sustainability efforts?
2. What are positive examples of linkages among agencies that have promoted urban sustainability?
3. How can the Federal government facilitate improvement in linkages, integration and synergy to promote urban sustainability?
4. What research, development and demonstration projects are needed to achieve this goal?

FIFTH COMMITTEE MEETING
July 16-19, 2012
J. Erik Jonsson Woods Hole Center
314 Quissett Ave Falmouth, MA 02543

This meeting was closed in its entirely.

SIXTH COMMITTEE MEETING
October 11-12, 2012
Keck Center of the National Academies
500 Fifth St NW Washington, DC 20001

October 11, 2012

9:00 am	*Welcome and Introductions* Dr. Thomas Graedel, Committee Chair, Yale University
9:15 am	*Forging Interagency Linkages on Sustainability: Federal Emergency Management Agency's Perspective* David Kaufman, Director, Office of Policy and Program Analysis, Federal Emergency Management Agency

9:45 am   *Forging Interagency Linkages on Sustainability: Department of Energy's Perspective*
                 Mark Gilbertson, Deputy Assistant Secretary for Site Restoration, Office of Environmental Management, U.S. Department of Energy

10:15 am   BREAK

10:30 am   *Public Administration Perspectives on Forging Interagency Linkages on Sustainability*
                 DeWitt John, Thomas F. Shannon Distinguished Lecturer in Environmental Studies, Bowdoin College

11:00 am   *Collaborative Governance and Sustainability*
                 Kirk Emerson, Professor of Practice in Collaborative Governance, University of Arizona

11:30 am   **Open Microphone Session: Brief Comments from Interested Parties**
                 Comments will be limited to 5 minutes, if you would like to address the committee please send an email to sustainability@nas.edu

12:20 pm   *Meeting Wrap-up*
                 Thomas Graedel, Committee Chair, Yale University

12:30 pm   Open Session Adjourns